T0395164

Artificial Intelligence in the Gulf

Elie Azar · Anthony N. Haddad
Editors

Artificial Intelligence in the Gulf

Challenges and Opportunities

Editors
Elie Azar
Khalifa University of Science and Technology
Abu Dhabi, United Arab Emirates

Anthony N. Haddad
Amazon
Dubai, United Arab Emirates

ISBN 978-981-16-0770-7 ISBN 978-981-16-0771-4 (eBook)
https://doi.org/10.1007/978-981-16-0771-4

Cover illustration: @Buena Vista Images/GettyImages

This Palgrave Macmillan imprint is published by the registered company Springer Nature Singapore Pte Ltd.
The registered company address is: 152 Beach Road, #21-01/04 Gateway East, Singapore 189721, Singapore

Acknowledgments

This volume is based on a selection of papers presented at the "Artificial Intelligence in the Gulf: Prospects and Challenges" workshop, which was held on July 15–18, 2019 at the University of Cambridge in the United Kingdom. The co-editors would like to extend a warm thank you to all participants of the workshop, who convened over three days to present academic papers and discuss the prospects of Artificial Intelligence in the six Gulf Cooperation Council member states. The discussions held during the event greatly enriched the chapters of this volume and helped the co-editors and authors together define a number of overarching themes, challenges, and solutions relating to the topic.

The co-editors are also deeply grateful to the organizers of the 2019 Gulf Research Meeting of the Gulf Research Centre Cambridge, under the auspices of which the workshop took place. In particular, we would like to thank Dr. Abdulaziz Sager, Chairman and Founder of the Gulf Research Centre (GRC), Dr. Oskar Ziemelis, Director of Cooperation at the GRC, Dr. Christian Kokh, Director of the GRC Foundation Geneva, as well as Ms. Aileen Byrne and Ms. Sanya Kapasi at the GRC, for their tireless work and support.

United Arab Emirates
June 2021

Elie Azar
Anthony N. Haddad

Contents

List of Contributors

Helen Abadzi University of Texas, Arlington, TX, USA

Aleksandar Abu Samra Khalifa University of Science and Technology, Abu Dhabi, UAE

Hesham Al-Ammal University of Bahrain, Zallaq, Bahrain

Ali al-Azzawi Smart Dubai, Dubai, UAE

Maan Aljawder University of Bahrain, Zallaq, Bahrain

Muhammad Ashfaq IU International University of Applied Sciences, Bad Honnef, Germany

Usman Ayub COMSATS University Islamabad, Islamabad, Pakistan

Elie Azar Khalifa University of Science and Technology, Abu Dhabi, UAE

Ahmed Badran Department of International Affairs, Qatar University, Doha, Qatar

Sahar ElAsad Regional Center for Educational Planning (RCEP), UNESCO, Dubai, UAE

Anthony N. Haddad Amazon, Dubai, UAE

Victoria Heath Montreal AI Ethics Institute, Toronto, ON, Canada

Toufic Mezher Khalifa University of Science and Technology, Abu Dhabi, UAE

Robert Mogielnicki The Arab Gulf States Institute, Washington, DC, USA

Mohammed Sharfi Independent Researcher and Senior Consultant, Ministry of Foreign Affairs, Doha, Qatar

Julia Singer Ludwig-Maximilians-University Munich, Munich, Germany

List of Figures

Chapter 10

Chapter 12

Chapter 13

List of Tables

Chapter 3

Chapter 6

Chapter 8

Chapter 10

Chapter 11

Chapter 12

PART I

Introduction

CHAPTER 1

An Introduction to AI in the GCC

Elie Azar and Anthony N. Haddad

1 Introduction

1.1 Background

Although the term Artificial Intelligence (AI) has appeared in literature as early as the mid-1950s (Grudin 2008), there still is some confusion with regard to its exact definition and context. The very early developments of AI were driven by the gaming industry, particularly to develop a computer program that could play chess (Newborn 2003). In 1997, nearly a half century later, an IBM supercomputer called DEEP BLUE won in a 6-game chess match against Russian Grandmaster Gary Kasparov, making the world chess champion the first to lose a match to a computer under standard time controls, and demonstrating the success of AI and its ability to outperform the human brain at a given task (Campbell et al. 2002).

E. Azar (✉)
Khalifa University of Science and Technology, Abu Dhabi, UAE
e-mail: elie.azar@ku.ac.ae

A. N. Haddad
Amazon, Dubai, UAE
e-mail: ahaddad@amazon.com

E. Azar and A. N. Haddad (eds.), *Artificial Intelligence in the Gulf*,
https://doi.org/10.1007/978-981-16-0771-4_1

Fast forward two decades, AI has now evolved from being driven by the development of computer games to complicated systems that are required for the functioning of our daily lives. As defined in a report by PricewaterhouseCoopers (PwC) (PwC 2018), "AI is a collective term for computer systems that can sense their environment, think, learn, and take action in response to what they are sensing and their objectives." AI systems can have various levels of autonomy and adaptiveness, ranging from systems that assist humans to perform tasks more efficiently to ones with automated decision-making processes without any human intervention (PwC 2018).

AI applications have gained a significant momentum in recent years across various sectors such as health care (e.g., medical data analysis and telemedicine), retail (e.g., product recommendations), manufacturing (e.g., assembly line automation), infrastructure (e.g., smart cities and self-driving cars), finance and banking (e.g., fraud detection), and education (e.g., personalized learning).

1.2 *Prospects*

AI is a disruptive technology in today's global economy. Beyond its implications on how people live and work, it has tremendous economic potential that remains untapped. With an incremental contribution estimated at US$15.7 trillion by 2030, this is more than the output of China and India combined (PwC 2017).

Given the nascent state of AI across countries, sectors, and individual businesses worldwide, emerging-market actors may have a key opportunity to leapfrog their more developed counterparts. The countries of the Gulf Cooperative Council (GCC), led by the Kingdom of Saudi Arabia (KSA) and the United Arab Emirates (UAE), have taken note. Compelled by the need to diversify their economies away from oil, the GCC states are actively pursuing knowledge-based economies. Moreover, the opportunity is significant, where among the GCC states, KSA is expected to benefit the most from the move toward AI with an expected contribution of US$135.2 billion toward its economy. The UAE, in turn, will benefit from a contribution of US$96.0 billion, while the remaining four countries are expected to share a total growth of US$45.9 billion. In relative terms to the GDP of each country, the AI contribution to the UAE economy will be the highest with 14% of its GDP (PwC 2018).

As a consequence, these countries have taken a progressive stance toward AI investment and adoption. In the backdrop of its Saudi Vision 2030 national transformation strategy, which has digital transformation identified as a key enabler (KSA Government 2016), the Kingdom famously granted Saudi citizenship to a female AI humanoid robot ("Sophia") at its first "Davos in the Desert" investment conference (Bloomberg 2018). More recently, in August 2019, it issued a royal decree to establish a National Authority for Data and AI, and hosted in October 2020 a virtual Global AI Summit (originally scheduled for May 2020 in Riyadh but delayed due to the coronavirus pandemic), touted to be the largest AI meetup of its kind, under the patronage of Crown Prince Mohammad bin Salman. Examples from the UAE include launching a national AI strategy (UAE Government 2017), launching an autonomous transportation strategy (Dubai Future Foundation 2017), appointing the world's first minister for AI (UAE Cabinet 2017), in addition to other initiatives such as establishing an AI university—with full-fledged Masters and PhD programs—based out of the country's capital, Abu Dhabi (Emirates News Agency 2019).

In response to the spread of the coronavirus, GCC countries have found important practical applications in AI technologies to help slow its spread. Enacting some of the world's strictest measures, including suspending passenger flights and imposing lockdowns with curfews, these countries have also turned toward speed cameras, drones, and robots to enforce social distancing and limited movement of citizens. In the UAE, Dubai Police uses a program called "Oyoon" which through a network of cameras in the city uses facial, voice, and license plate recognition to determine if a resident is employed in a vital sector or in possession of a valid permit to leave their home or business (CNBC 2020). Additionally, the city's Road and Transport Authority (RTA) implemented computer vision and machine learning algorithms to scan hundreds of thousands of hours of video footage to detect and report violators of preventative measures like proper wearing of masks, and proper distance between passenger and driver (Emirates News Agency 2020). Government authorities have also developed and heavily enforced the use of mobile phone apps, like Tawakkalna and Tabaud in Saudi Arabia, and BeAware in Bahrain, to facilitate location-based contact tracing that monitors those who have tested positive for the virus and to try to limit their exposure to the population (The National 2020).

While headline-grabbing in nature, these initiatives are not isolated, one-off events: They are part of serious attempts by the GCC states—led by KSA and the UAE—to take a progressive and deliberate step toward AI investment and adoption as an engine for growth. In so doing, they seek to chart a path toward transitioning to a knowledge-based economy, with the intent to compete with the world (and each other).

Yet in spite of this heightened level of GCC interest and investment in AI, there are surprisingly no serious efforts to date to pull together scholarship on the adoption challenges that exist, the implications on society that these technologies may have, and the regulatory requirements that need to be in place. Related efforts have typically been segregated and remain limited in comparison with other growing economies such as China and India. Important barriers are expected to hinder the full deployment of AI in the GCC. These include—but are not limited to—locally-generated knowledge, economic barriers, social risks, and rigid institutional and policy structures. As such, important questions arise and need to be further studied:

1. Are the Gulf countries on track to catch-up with the global push toward AI?
2. Are the current physical and institutional infrastructures ready for such transformation?
3. Is local talent available and ready?
4. Is the GCC economy ready for such job market evolution?
5. Will GCC countries remain adopters of AI knowledge and technology or will they evolve to become effective developers and contributors at the local and national stages?

1.3 *Objectives, Scope, and Target Audience*

This book fills an important gap in the literature, with the first broad reflection on the challenges, opportunities, and implications of AI in the GCC. Unique results and insights are derived through case studies from diverse disciplines, including engineering, policymaking and governance, economics, social science, and data science. Particularly related to the "soft" science disciplines, we make some unexplored yet topical contributions to the literature, with a focus on the GCC (but by no means limited to it), including: AI and implications for women, Islamic schools of thought on AI, and the power of AI to help deliver well-being and happiness in cities and urban spaces. Finally, the book also provides readers with a synthesis of ideas, lessons learned, and a path forward based on the diverse content of the chapters. As such, we seek to have this be a foundational text in what promises to be a growing area of research not only in AI in the GCC, but of the contributions to reducing the existing digital "North–South divide."

The book caters to the educated non-specialist with interest in AI, targeting a wide audience including professionals, academics, government officials, policymakers, entrepreneurs, and non-governmental organizations. Given the gap in the literature on AI in the Gulf regional context, the book is deliberately broad and diverse with chapters encompassing multiple disciplines. As such, the volume is not intended to serve as a textbook to undergraduate or graduate courses, though we see chapters or parts (e.g., Part 3—Society, Utopia and Dystopia) being used in courses. Finally, all of the chapters present original methods, results, and analyses, which we anticipate will generate significant interest and citations in future scholarly work on the topic.

As detailed in the following chapter, the book comprises 13 chapters organized along three main themes pertaining to AI in the GCC: (1) Data, Governance and Regulations; (2) Existing Opportunities and Sectoral Applications; and (3) Society, Utopia and Dystopia.

The core of our contributions comes from a workshop held during the annual Gulf Research Meeting (GRM), which took place on July 15–18, 2019 at Cambridge University in the United Kingdom. The workshop was entitled "Artificial Intelligence in the Gulf Cooperative Council (GCC) countries: Opportunities and Challenges" and comprised of academics, practitioners, and policymakers of AI-related disciplines from the six GCC countries (Saudi Arabia, United Arab Emirates, Kuwait, Qatar, Bahrain, and Oman).

References

Bloomberg. (2018). *Saudi Arabia Gives Citizenship to a Robot*. News article. https://www.bloomberg.com/news/articles/2017-10-26/saudi-arabia-gives-citizenship-to-a-robot-claims-global-first.

Campbell, M., Hoane Jr., A. J., & Hsu, F. H. (2002). Deep Blue. *Artificial Intelligence*, 134 (1–2), 57–83.

CNBC. (2020). *Some Countries in the Middle East are Using Artificial Intelligence to Fight the Coronavirus Pandemic*. News article. https://www.cnbc.com/2020/04/16/countries-in-the-middle-east-are-using-ai-to-fight-coronavirus.html.

Dubai Future Foundation. (2017). Dubai Autonomous Transportation Strategy. *Dubai Future Foundation*, Dubai, UAE.

Emirates News Agency. (2019). Abu Dhabi Announces Establishment of the Mohamed bin Zayed University of Artificial Intelligence. News article. *Emirates News Agency*, Abu Dhabi, UAE. https://wam.ae/en/details/1395302795116.

Emirates News Agency. (2020). RTA Employs AI in Taxis to Curb Spread of COVID-19. News article. *Emirates News Agency*, Dubai, UAE. https://wam.ae/en/details/1395302848858.

Grudin, J. (2008). A Moving Target: The Evolution of HCI. In *The Human-Computer Interaction Handbook—Fundamentals, Evolving Technologies, and Emerging Applications*. Lawrence Erlbaum Associates, Taylor & Francis Group, New York, NY.

KSA Government. (2016). KSA Vision 2030. https://vision2030.gov.sa/en. *KSA Government*, Riyadh, KSA.

The National. (2020). *AI Helped Limit Spread of Covid-19 in the Gulf, Experts Hear*. News article. https://www.thenational.ae/uae/health/ai-helped-limit-spread-of-covid-19-in-the-gulf-experts-hear-1.1063052.

Newborn, M. (2003). *Deep Blue: An Artificial Intelligence Milestone*. Springer, New York, NY.

PricewaterhouseCoopers (PwC). 2017. *Sizing the Prize: What's the Real Value of AI for your Business and How Can You Capitalize?* PwC, London, UK.

PricewaterhouseCoopers (PwC). (2018). *The Potential Impact of AI in the Middle East*. PwC, London, UK.

UAE Cabinet. (2017). His Excellency Omar Bin Sultan Al Olama. *UAE Cabinet*, Abu Dhabi, UAE. https://uaecabinet.ae/en/details/cabinet-members/his-excellency-omar-bin-sultan-al-olama.

UAE Government. (2017). UAE Strategy for Artificial Intelligence. *UAE Government*, Abu Dhabi, UAE.

CHAPTER 2

Framework of Study and Book Organization

Elie Azar and Anthony N. Haddad

1 Book organization

The book consists of five main parts. Part I: *Introduction*—which is comprised of two chapters—introduces the readers to the concept of Artificial Intelligence with a historical review both from the international and local GCC contexts, followed by a mapping of the current gaps in the literature motivating the need for the current volume. Part II: *Data, Governance & Regulations*—which includes three chapters—presents case studies on the supporting milieu needed to leverage advancements in the context of multiple GCC cities. Part III: *Existing Opportunities & Sectoral Applications*—which is composed of three chapters—covers case studies and applications of AI initiatives in multiple sectors, such as finance, economics and healthcare. Part IV: *Society, Utopia and Dystopia*—which includes four chapters—offers a unique social perspective of AI, with insights from studies on gender, religion, psychology and happiness. Part

E. Azar (✉)
Khalifa University of Science and Technology, Abu Dhabi, UAE
e-mail: elie.azar@ku.ac.ae

A. N. Haddad
Amazon, Dubai, UAE

E. Azar and A. N. Haddad (eds.), *Artificial Intelligence in the Gulf*,
https://doi.org/10.1007/978-981-16-0771-4_2

VI: *Conclusion*—which presents an overview of the topics covered in the various chapters—highlights key findings and directions for future research. The following paragraphs detail each of the chapters presented in the book starting from the chapters of Part II, which follow the current introductory chapters.

2 Summary of Chapters

In Chapter 3, "*Public Sector Data for Academic Research: the case of the UAE*", the authors explore the current procedures, barriers and possible solutions for disclosing Public Sector Information (PSI) for academic research in the UAE. The authors then propose a solutions framework validated with public officials to better facilitate data sharing in the country, which is an important prerequisite for research on data-driven fields, such as AI.

In Chapter 4, "*Strategy for Artificial Intelligence in Bahrain: Challenges and Opportunities*", an AI strategy is presented for the Kingdom of Bahrain following a multi-stakeholder analysis. The strategy is discussed from different perspectives, including the availability of human capital and the support from economic and financial institutions, while also aligning it with the national strategy plans and priorities.

In Chapter 5, "*Thoughts and Reflections on the Case of Qatar: Should Artificial Intelligence be Regulated?*", the author tackles a fundamental question about the role of government in regulating AI in the light of the growing ethical, legal and security concerns of this technology. Based on insights from the AI policy community in Qatar, the author calls for a regulatory intervention from governments in order to strike a balance between potential benefits and the expected threats from AI systems and applications.

In Chapter 6, "*Knowledge, Attitude, and Perceptions of Financial Industry Employees toward AI in the GCC Region*", the authors investigate the knowledge, attitude and perceptions of professionals working in the financial services industry in all six GCC countries. A statistical analysis of data collected from 157 professionals shows an overwhelming familiarity of the respondents with AI in the business and financial sector context. The findings also highlight important concerns regarding ethical, security and data privacy issues.

In Chapter 7, "*The GCC and Global Health Diplomacy: The New Drive towards Artificial Intelligence*", AI is studied in the context of its impact on the healthcare sector of GCC states and the emerging concept of Global Health Diplomacy (GHD). The premise of this chapter is that the financial resources provided by the GCC for AI applications in healthcare will positively contribute to GHD. In parallel, the need to engage in global partnership and collaboration is highlighted to effectively leverage the power of AI in the GCC health sectors.

In Chapter 8, "*Free Zones in Dubai: Accelerators for Artificial Intelligence in the Gulf*", a unique analysis is presented on how economic free zones can serve as useful vehicles for the effective implementation of AI technologies in Gulf Arab economies. Taking Dubai as a case study, the author argues that an alignment between AI innovation and ongoing free zone developmental processes provides an amendable and sustainable environment that maximizes the impact of AI services and applications.

In Chapter 9, "*AI & Well-Being: Can AI make you happy in the city*", the manifestation of AI technologies and applications is mapped to the way that happiness and well-being are understood in the various contexts of people's lives. Drawing on examples from Dubai, and other global initiatives, the author explores the utility of AI towards happier lives from various perspectives, offering directions for further work to ensure a wider coverage for all aspects of well-being.

In Chapter 10, "*Women and the Fourth Industrial Revolution: An Examination of the UAE's National AI Strategy*", the participation of women in the development, deployment and governance of AI is studied. Using an extensive review of academic literature, news articles and data from government and intergovernmental organizations, the author argues that women's inclusion in AI is a significant factor in ensuring the successful development, deployment and governance of AI. A potential strategy to include women in national AI strategies is also presented.

In Chapter 11, "*The Art and Science of User Exploitation: AI in the UAE and Beyond*", the authors explore the misuse of AI by Internet companies to exploit vulnerabilities in human psychology and influence users' views on matters ranging from political stands to promoting product sales. The chapter provides unique insights on the neuroscience behind the persuasion tactics used by these businesses, the strides and challenges of AI use globally and in the Gulf region, as well as the implications of AI across different disciplines.

In Chapter 12, "*Fatwas from Islamweb.net on Robotics and Artificial Intelligence*", the author evaluates the perceptions of robotics and AI from an Islamic perspective. For that purpose, 14 Arabic and English

"Fatwas" are studied, which are legal opinions expressed by a Muslim scholar or people with expertise in Islamic Law. The findings show that scholars (1) have a fairly clear stance on the treatment of robotics but not on AI; (2) they also do not show concerns with these technologies potentially harming humans; and (3) avoid difficult issues such as the impacts of developing strong AI.

Finally, in Chapter 13, "*Outlook for the future of AI in the Gulf*", the book is concluded with an overview of the topics and case studies that were covered in the various chapters. The chapter includes a discussion of the key findings, insights learned, recommendations on how to address the challenges towards a more effective implementation of and advancement of AI technologies in the GCC, with directions for future research on the topic.

PART II

Data, Governance and Regulations

CHAPTER 3

Public Sector Data for Academic Research: The Case of the UAE

Aleksandar Abu Samra, Toufic Mezher, and Elie Azar

1 Introduction

Access to government data is one of the most important components of economic and social development in society (Heeks, 2002). It provides value to citizens and organizations, promotes public participation, improves the decision-making process, and fosters creative and innovative solutions to contemporary problems (Janssen, 2011). More specifically, in academic circles, acquiring information is critical to conducting research and expanding scientific solutions (Mopas & Turnbull, 2011). However, this process oftentimes faces many challenges, especially when information lies behind the gates of government authorities. Academic researchers often rely on their personal negotiation skills to get the information needed because going through formal channels proves to be hectic and time-consuming. Resistance to data sharing persists worldwide because of institutional challenges rather than technical ones, "people"

A. Abu Samra · T. Mezher · E. Azar (✉)
Khalifa University of Science and Technology, Abu Dhabi, UAE
e-mail: elie.azar@ku.ac.ae

E. Azar and A. N. Haddad (eds.), *Artificial Intelligence in the Gulf*,
https://doi.org/10.1007/978-981-16-0771-4_3

challenges (Harvey & Tulloch, 2006). The problems of hardly approachable and unresponsive authorities are widely recognized and actively addressed by open government and transparency initiatives, but the situation varies significantly between different countries.

Thanks to rapid technological improvements, data on many different types of information has become abundant, easy to collect, and inexpensive to store (Lane et al., 2014). As key players in the infastructure sectors are typically government entities and public-private partnerships, relevant data naturally falls under the public domain. Acknowledging the value of collected information, but also limitations to coping with its vastness, it is not surprising that governments around the world are introducing new data-sharing policies to engage other stakeholders toward understanding and getting insights out of it (Einav & Levin, 2013). Freedom of Information Acts and Open Data movements are on the rise, and now more than ever, datasets are possible to combine across borders and disciplines, further accelerating the societal progress.

In the United Arab Emirates (UAE), recent investments into higher education hubs and a rapidly evolving research ecosystem aim to place the country as a global leader in research and innovation (Byat & Sultan, 2014; Al-Hammadi et al., 2010), particularly in the field of Artificial Intelligence (AI). This is confirmed by the number and scale of AI initiatives that are being undertaken, such as the launching of the national AI strategy or the new Mohamed bin Zayed University of Artificial Intelligence (UAE Government, 2017; Emirates News Agency, 2019).

However, despite these advancements, data sharing policies remain limited both in terms of Open Data (World Wide Web Foundation, 2015), and more rudimental Right to Information laws (World Justice Project, 2015). For this reason, access to Public Sector Information (PSI) is burdensome to scientists, a condition that was shown to significantly slow down innovation and the research potential in many countries (Alani et al., 2007; Arzberger et al., 2004; Commission of the European Communities, 1999).

There is a great need to promote data sharing and collaboration between government and academia in the UAE, especially for data-intensive fields, such as AI. Although there are many studies, globally and from the region, that assess the state of openness through international rankings, there is a lack of studies exploring this problem in more depth and focusing on specific challenges and potential solutions.

The main goal of this chapter is to comprehensively evaluate and understand the current state of sharing PSI in the UAE, as well as propose feasible solutions for improving it. For achieving that goal, objectives were split into the following segments:

- Understand what logistical, communicational, legal, and personal challenges researchers encounter in the process of obtaining PSI necessary for their studies.
- Understand what protocols are followed by the UAE authorities when approached with data requests and detect the key barriers to sharing government data with third-parties, particularly academia.
- Understand the latest international practices used to promote the sharing of PSI and possible implementation challenges in the UAE.
- Propose recommendations to modify international practices to address locally discovered challenges and broader understanding of the local context.
- Combine the developed recommendations to propose a solution framework consisting of policies, strategy, and action timeline to introduce improved data sharing in the UAE.
- Validate proposed solutions with local authorities and experts in the field.

2 Literature Review

2.1 Challenges to Data Sharing

Challenges to freedom of information are certainly very common in both developed and developing parts of the world (Banisar, 2006), with notable debates happening in the fields of access to citizen data, ethics of data sharing, and in the regulatory fields (Cohen et al., 2014; LAPSI 2.0, 2014).

From a user's perspective, challenges to government information access vary by demographics and were shown to be experienced in Research and Development (R&D) (Beniston et al., 2012; Mopas & Turnbull, 2011), by journalists (Garrison, 2000), in inter-organizational sharing (Fedorowicz et al., 2010), and at the international scale (Harris & Browning, 2013).

Sharing knowledge between individuals and organizations is a complex issue, influenced by anything from personal and psychological incentives to institutional procedures and laws (Yang & Maxwell, 2011). At individual level, information can be perceived as property and therefore becomes an asset worth protecting for increasing one's status and influence, or used as a tool for securing personal power inside organizations (Ardichvili et al., 2003; Constant et al., 1994; Marks et al., 2008; Willem & Buelens, 2007). Across organizations, barriers to sharing information can be the time and effort needed to provide it, as well as exposure to criticism, especially if done without receiving proper recognition (Ardichvili et al., 2003; Cress et al., 2006). Furthermore, due to limited resources, organizations naturally prioritize immediate issues over long-term benefits to information exchange (Landsbergen & Wolken, 2001; Zhang & Dawes, 2006). Inter-organizational sharing of knowledge encounters further problems due to different origins, values, and cultures of participating organizations, often invoking complexity, misunderstanding, distrust, or even competition (Drake et al., 2004).

2.2 *Open Data and Open Government Data (OGD)*

Open Data, defined as data that is "free to use, modify, and share by anyone for any purpose" (Open Knowledge Foundation, 2015), enjoys increased attention going hand in hand with the rapid development of technology and information distribution. Government authorities, although in possession of much-valued data on society, do not have resources to process it alone, and need to outsource the initiation of societal challenges to their citizens (Eckartz et al., 2014). When addressing strictly PSI, the term Open Government Data (OGD) is commonly used, annotated by eight original principles stating the data should be complete, granular, timely, accessible, machine-processable, non-discriminatory, non-proprietary, and license-free (Tauberer, 2007).

Still, Open Data has many challenges on its own. Publishing structured datasets is new for local governments and, therefore, causes confusion on how data should be released (Conradie & Choenni, 2014). Some of the dominant barriers include data quality, ownership and trust issues, privacy, and economic impediments. To unlock its potential, data needs to be governed (Cheong & Chang, 2007; Weill & Ross, 2004), but there is a lack of tools for sharing, data definitions are conflicting, and privacy concerns are many (Zhang et al., 2005). Practical frameworks have been developed to guide through these barriers, but on the local levels, there is oftentimes an absence of understanding of the added value of such change (Kassen, 2013). Table 1 presents an extended list of the most relevant documents used for understanding international practices in this study.

2.3 Information Sharing Trends in UAE

The UAE national goals are clearly set on the direction toward a knowledge-based economy and society. The UAE Vision 2021 (Prime Minister's Office, 2010) announced a national era of competitive knowledge economy based on entrepreneurship and innovation. Furthermore, the Abu Dhabi Department of Economic Development (2009) has acknowledged that the knowledge-based economy is not merely a new branch of economy, but rather an entire transition to an economy based on information. However, as data is the key driver of such change, it is important to address the current gaps to harnessing its full potential. In the latest Open Data Barometer report (World Wide Web Foundation, 2015), the country scored low compared to its broader state of prosperity. An equivalent to a "Freedom of Information" act is still not available, and the UAE generally ranks low in Open Government Index by the World Justice Project (2015). Finally, a yearly review of datasets published by government agencies in the UAE carried out by the Global Open Data Index (2015) ranked the UAE at the bottom 20% of all reviewed countries concerning various information availability and structure criteria.

In recent years, Dubai promised to bring positive progress to this area (Government of Dubai Media Office, 2015). However, few studies have already tackled the current state of data "openness" in the UAE, looking at information present on the official government Web sites (AlAnazi & Chatfield, 2012) or directly interviewing the authorities (Elbadawi, 2012). Still, little research efforts have been conducted on problems that emerge when governmental data is directly requested for academic use, exploring issues of the current government–academia relationship over data in the UAE, or on how international trends can be applied to the local context.

This chapter aims at closing this gap by interviewing both academicians and public officials in the UAE in order to investigate how government authorities approach data requests by academia, and what are their perceptions of the trending shifts toward Open Data. A proposed solution

Table 1 Resources of best practices in OGD

Source Type	*Reference*
International Organizations	Open Government Data Toolkit (World Bank, 2016)
	Open Data Handbook (Open Knowledge Foundation, 2016)
	Open Data: Measuring What Matters (Hadjigeorge, 2016)
	Open Data Barometer Global Report (World Wide Web Foundation, 2015)
	International Open Data Charter (Group of Eight, 2015)
	Guidelines for Open Data Policies (Sunlight Foundation, 2014)
Federal Governments	Open Data Policy—Managing Information as an Asset (Project Open Data, 2016)
	Open Data in the G8: A Review of Progress on the Open Data Charter (Castro & Korte, 2015)
	Supporting the National Information Infrastructure (Public Data Group, 2014)
	Open Data Strategy 2014–2016 (Department for Business Innovation & Skills, 2014)
	Canada's Action Plan on Open Government 2014–2016 (Treasury Board of Canada Secretariat, 2014)
	Open Government Plan (General Services Administration, 2012)
City and County Governments	LA Open Data Policy and Playbook (Los Angeles Data Team, 2015)
	Open Data Guide by the City of Philadelphia (Headd, 2014)
	Open Data in San Francisco: Institutionalizing an Initiative (Bonaguro, 2014)
	Open Data Implementation Plan (Montgomery County Government, 2014)
	Open Data Policy—Implementation Plan (City of San Diego, 2014)
	Open Data Policy and Technical Standards Manual (Bloomberg & Merchant, 2012)
	NYC OpenData Technical Standards Manual (NYC OpenData, 2012)

Source Type	*Reference*
Research Papers and Case Studies	Planning and designing open government data programs: An ecosystem approach (Dawes et al., 2016) A systematic review of open government data initiatives (Attard et al., 2015) Open Data in the Legislature: The Case of São Paulo City Council (Matheus & Ribeiro, 2014) Open Data Ireland: Best Practice Handbook (Lee et al., 2014) Open Government Data: Towards Empirical Analysis of Open Government Data Initiatives (Ubaldi, 2013) Open data: Emerging trends, issues and best practices-a research project about openness of public data in EU local administration (Fioretti, 2012) Open data: an international comparison of strategies (Huijboom et al., 2011)

framework is then presented to guide efforts toward implementing such trends while staying consistent with the current regulatory limitations and practices in place.

3 Methodology

The methodology used for conducting the interviews in this study was split into five phases: (1) interview design (2) data collection (3) data analysis and identification of barriers (4) understanding international practices and their challenges, and (5) proposing a comprehensive solution framework. Figure 1 visualizes details of this methodology:

3.1 Interview Design

The aim of the interview design phase was to provide a tool to collect information about the process of obtaining datasets from the public sector in the UAE, from both the perspective of academicians (who request the datasets) and public officials (responsible for processing those requests). To capture the full spectrum of subjects' experiences, a face-to-face qualitative interview approach was used following best practices in interview protocols (Seidman, 2013; Turner III, 2010; Weiss, 1994).

The interview with academicians had the objective of identifying logistical, communicational, legal, personal, and any other challenges encountered in obtaining PSI needed in their studies. The interview also explored the nature of the PSI requested, as well as personal opinions on how the researchers thought the system might be improved. The interview was split into four sections:

1. *Getting to Know the Interviewee*—To briefly give a context of the subject's background and ongoing work.
2. *Project-Related Details*—Aimed at understanding why subjects need data from public authorities, the complete process of requesting it, and the related barriers in obtaining it.
3. *Thoughts on Possible Improvements*—Subjects are asked to give their opinions on the current state of data availability in the UAE, and steps toward improving it.
4. *Closing the Interview*—Lastly, the subjects are asked to refer to other people that experienced similar problems and give general feedback on the interview process.

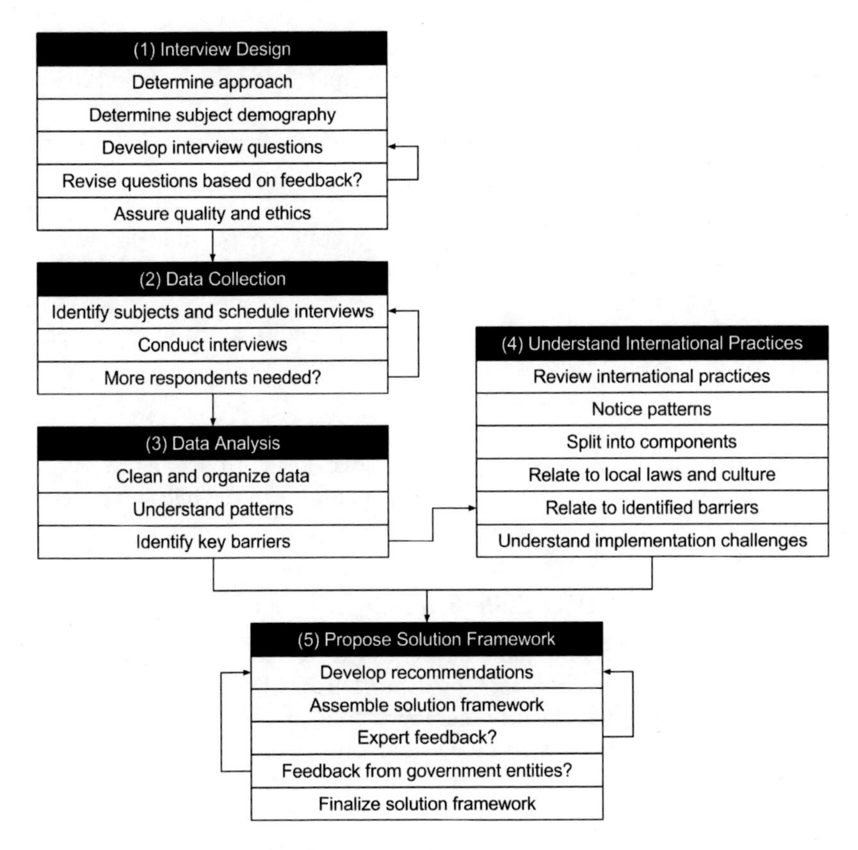

Fig. 1 Methodology development

The interview with public officials had the objectives of understanding protocols followed by the UAE authorities when approached with data requests from academia, the concerns these requests might cause, and personal opinions on how these concerns could be mitigated. The interview was split into five sections:

1. *Getting to Know the Interviewee*—A brief introduction to the subject's position in his or her organization.

2. *Data Collection Processes*—Understanding what type of data the authority collects or possesses, the formats used, and their access hierarchy.
3. *Infrastructure for Open Data*—Understanding familiarity with Open Data, current data sharing protocols in place, and barriers to a more open approach.
4. *Thoughts on Openness*—Collecting opinions on possible improvements in data sharing practices from both government and from academia sides.
5. *Closing the Interview*—Final thoughts, feedback from the interviewee, and references for further interviews.

After developing the first versions of both interviews, test-runs were performed with subjects that agreed to give in-depth feedback. As a result, some questions were shortened and made clearer, while some were identified as redundant and therefore removed.

3.2 Data Collection and Scope of Work

Twenty-two researchers in academia were interviewed across ten research institutions across the Emirates of Abu Dhabi, Dubai, Sharjah, and Ajman. This included 17 faculty members, four graduate students (three Doctoral and one Master's student), and one research engineer. Three participants were three females and 19 males. Initial subjects were identified by word of mouth recommendation from experts in the field, followed by referrals acquired by explicitly asking subjects for further contact suggestions. Interviews were scheduled by sending a short introductory email directly, asking for a 20-minute interview benefiting the study. If the respondents seemed to be busy, they kept things to the point and within the announced interview time, but if the respondents had extra time and wanted to contribute more, interviews continued for up to an hour.

Twelve people were interviewed across nine public authorities in the UAE and one consultancy company involved in local e-government efforts. Contact details of potential subjects were initially acquired by reaching their respective departments through information found online, after which a call, email, or snail-mail letter to the department would follow. Later on, the "snowball sampling" approach was used, where the investigator asked already interviewed subjects to refer him to relevant people in other public authorities. The investigator would then email or call the new contacts, briefly introduce the study, and ask to schedule an interview.

3.3 *Data Analysis and Identification of Barriers*

In analyzing notes collected during the interviews, a general inductive approach was used for analyzing qualitative data (Thomas, 2006). In the results section, findings are presented as objectively as possible, basing them solely on interviewees' responses. Key barriers to data sharing were then identified from both the perspectives of academic researchers and government officials.

3.4 *Understand Challenges to Implementing Best Practices*

This phase of the methodology first consists of thoroughly reviewing best practices in Open Government Data (OGD) from a multitude of sources including international Open Data organizations, policy documents and guidebooks from federal and city governments, as well as research papers and case studies reviewing and analyzing different OGD strategies. After the practices had been properly understood, recognized patterns were split into separate *Best Practices* for easier further analysis. Each major recognized *Best Practice* was related to the context of UAE culture and procedures in place, as well as the previous findings on the existing key barriers to data sharing. Arguments were made on why, if the UAE was to follow such international practices directly, challenges could be encountered and the progress potentially stalled.

3.5 *Propose Solution Framework*

Based on each recognized *Best Practice* of Open Data, and for every recognized challenge to its implementation in the UAE, a specific recommendation was proposed to address it. The adapted *Best Practices* were finally assembled in a solution framework, proposing how such a modified approach to Open Data could be successfully implemented in the Emirate of Abu Dhabi. The developed solution framework, which is detailed in the upcoming sections, consists of four layers; each layer incorporated relevant *Best Practices* and their respective developed recommendations, while again referring to barriers identified from the previously held interviews:

1. *Strategy*—Describing the proposed policy and actions performed by the entities responsible for its development;
2. *Execution*—Describing actions of the government entity proposed to put the developed strategy into action;
3. *Operation*—Describing responsibilities of government authorities that would participate in this solution by providing datasets to the system;
4. *Process*—Describing how the solution works from a user perspective.

To validate such a solution framework, the author held meetings with policy experts and people in the UAE government authority, some of whom also participated in the previous interviews. Based on the given feedback, and for each suggestion relevant to the scope of the study, recommendations were updated, and the framework reassembled. After the saturation point was reached, the proposed solution framework was considered to be in its final version.

4 Results and Analysis

4.1 Barriers to Accessing Public Sector Information

From the perspective of interviewed researchers, feelings about the current government–academia relationship in the UAE were mixed. In many cases, the researchers reported the willingness of public officials to help, but the attitude was never enough to lead to successful outcomes. Obstacles commonly faced by researchers include identifying who has the data and obtaining the requested information in a timely manner and in the right format. Lack of relevant laws and regulations caused data requests to appear unusual, requiring multiple in-person meetings with government officials to justify such requests. Furthermore, the researchers had to be affiliated with institutions that had pre-signed Memorandums of Understanding (MoUs) with respective government bodies, and finally, the researchers would have to sign Non-Disclosure Agreements (NDAs), preventing them from sharing data even with their colleagues.

The interviews within public authorities in the UAE revealed the complementing side of the narrative: The UAE authorities indeed have good intentions in supporting local academia and research, but the processes to attain such intentions with successful data exchange are unclear, and many obstacles exist toward a faster and more open approach.

Table 2 Reported barriers to PSI

Academia's perspective	*Public authorities' perspective*
Identifying Authorities with Needed Data	Absence of Data Sharing Laws and Regulations
Finding Contacts and Initiating Communication	Concerns about Quality, Completeness, and Availability of Data
Lack of Procedures	Security and Privacy Concerns
Paperwork	Misuse and Misinterpretation Concerns
Slow Response by Authorities	Lack of Incentives to Share Data
Unstructured and Undocumented Data	
Unfamiliarity with MoUs and NDAs	

The perception of interviewed respondents was that the current systems in place performed well. Obstacles to a more open approach to data sharing are many and include lack of laws and legislations, concerns about the quality of shared data, but also concerns on how the shared data would be used and represented. Employees within public authorities see no immediate benefits to sharing data and therefore lack direct incentives to optimize their processes. In summary, Table 2 presents the most commonly reported barriers by the interviewees.

4.2 *Adapting International Best Practices*

The following are summaries of main *Best Practices* of implementing Open Data on a government level, selected from subjectively recognized patterns in the international practices referenced in Table 1. In order to highlight the differences between the UAE and the countries who developed these best practices, each of the *Best Practices* is carefully studied, identifying its possible challenges to implementation in the local context. Finally, specific recommendations were developed to modify each *Best Practice* for successfully solving such challenges.

4.2.1 Best Practice #1: Data Sharing Principles

In short, Open Data is defined as: "*Data and content that can be freely used, modified, and shared by anyone for any purpose*" (Open Knowledge Foundation, 2015). Together with the eight underlying principles of OGD (Tauberer, 2007), Open Data is not fully compatible with the current practices of government authorities in the UAE. In the absence of laws on rights to information access, the UAE public authorities feel a significantly larger degree of direct "ownership" for such collected information. For a successful solution that would facilitate improved access to PSI, the political and cultural realities must first be fully understood. As the main issues of accessing PSI by researchers in the UAE are of communicational and procedural nature, local authorities should focus on improving access to data that exists already, in the form that is already kept, with permissions and usage rights that are already in place.

4.2.2 Best Practice #2: Government Structure Implementation

International practices to implementing OGD typically rely on appointing new, dedicated government bodies for its plan and execution. To avoid the logistical complexity and unnecessary risk such approach might cause, Abu Dhabi can decide for a more integrated approach: Abu Dhabi Systems & Information Centre (ADSIC) is leading the current information sharing in the capital Emirate, with a mission to "*[…] Deliver innovative digital services, standards and policies, while building the foundation for an ICT-Mature society*" (ADSIC, 2016). ADSIC has already implemented the Abu Dhabi Spatial Data Infrastructure (AD-SDI) program that focuses on sharing geospatial data among numerous public stakeholders in the Emirate. The government-to-government data communication they provide uses modern, sophisticated systems, with MoUs already in place. More than any currently existing public authority in Abu Dhabi, ADSIC has the know-how, the experience, and the government support to facilitate data exchange between different stakeholders of PSI. For those reasons, ADSIC could potentially form another internal body for executing the newly introduced data sharing practices.

4.2.3 Best Practice #3: Identifying Publishable Datasets

Various international practices for identifying publishable datasets focus on transparency of internal government processes, especially in countries funded by taxpayer's money; the emphasis on such information in the UAE is expected to cause unnecessary implementation difficulties. Therefore, the guidelines for deciding on datasets valuable for publishing should first exclude any data related to government expenses, internal processes, people accountable, or other similar information. Efforts saved by not publishing such data can largely be used for focusing on datasets of direct value to the local stakeholders. More specifically, combining approaches mentioned in the international practices, authorities should focus on publishing data directly benefiting research, innovation, and economy in the country.

4.2.4 *Best Practice #4: Prioritization Criteria*

Similarly, to Best Practice #3, international OGD practices often advise authorities to prioritize datasets that would increase transparency and accountability of their internal decisions and processes. Given the barriers listed in Table 1, priority cannot be fully put on the transparency issues, risking rejection by both policymakers and authorities themselves. As a compromise solution, this study recommends prioritizing the publication of data according to the UAE national goals and future visions.

4.2.5 *Best Practice #5: Data Portal and Access Rights*

A common practice included in OGD solutions is having a unifying web portal that serves as a structured, categorized, and searchable database of all "public" information published by relevant government authorities to date. Taking into account the current strict criteria for perceiving data as "public," the majority of the datasets would still fall under the "restricted public," or "non-public" category, therefore unavailable through the web platform provided. Consequently, a newly created portal must not focus solely on the collection and categorization of datasets (as typical Open Data portals do) but instead serve as a platform to facilitate communication between people and organizations in need of data on one side, and public authorities on the other. Because many of the datasets owned by the UAE authorities are considered "restricted," the portal should also route requests for such data to relevant authorities to resolve.

4.2.6 *Best Practice #6: Phasing Out Implementation*

A typical timeline for international OGD implementations often includes direct steps toward satisfying the end goal of any such implementation; such steps can be announcing an Open Data Policy, appointing an Open Data Board and a Chief Data Officer, or developing a unifying Open Data Portal. These implementation practices, although reasonable and not overly ambitious, would not be suitable for the context of UAE for one main reason: Such order of developments does not immediately address the most critical challenges of data access in the country, which is facilitation communication between data users and public authorities. To immediately address the critical challenges of PSI access in Abu

Dhabi, a three-phase approach is therefore recommended, as detailed in Sect. 4.3.3.

4.2.7 *Best Practice #7: Engaging Stakeholders*

To catalyze the use of published data, international OGD practices focus on stakeholder engagement with events targeting the general public. A major challenge to following such practice in the UAE is the assumption that the majority of datasets will still be restricted to only academia and other important stakeholders with special permission to use it. Consequentially, a more optimal approach would be to focus engagement efforts on important stakeholders, as currently done by ADSIC, proposed in *Best Practice #2.*

4.2.8 *Best Practice #8: KPIs and Accountability Practices*

KPIs measuring the performance of international OGD efforts mainly fall into three categories: (1) activity metrics (2) quality metrics, and (3) impact metrics. Such developed metrics do not measure the current most important challenges in the UAE—the general responsiveness of government authorities to data requests. As the major complaints in the current process of obtaining datasets from public authorities are of communicational and procedural nature, in addition, the metrics that target *Quantity*, *Quality*, and *Impact* of data, *Timeliness* of responses to data requests should be not only introduced but prioritized. Taking the time needed for successfully processing each request is critical for incentivizing authorities toward a more dedicated performance. Moreover, the metrics on *Quantity* and *Quality* should also include metrics specialized in tracking the performance of authorities' responses to data requests.

4.3 *Proposed Comprehensive Solution Framework*

A macro-sketch of the proposed comprehensive framework is provided in Fig. 2 for easier visualization and communication of proposed ideas to the reader. The framework combines developed *Best Practices* of relevant international practices, respective recommendations on how they can be better implemented in the UAE, and currently existing barriers identified after analyzing the performed interviews. On the left side of Fig. 2 are listed brief summaries of the investigated challenges to data sharing. Each is marked with a flag number to indicate what elements of the proposed framework play a role in addressing the existing problems. Within the

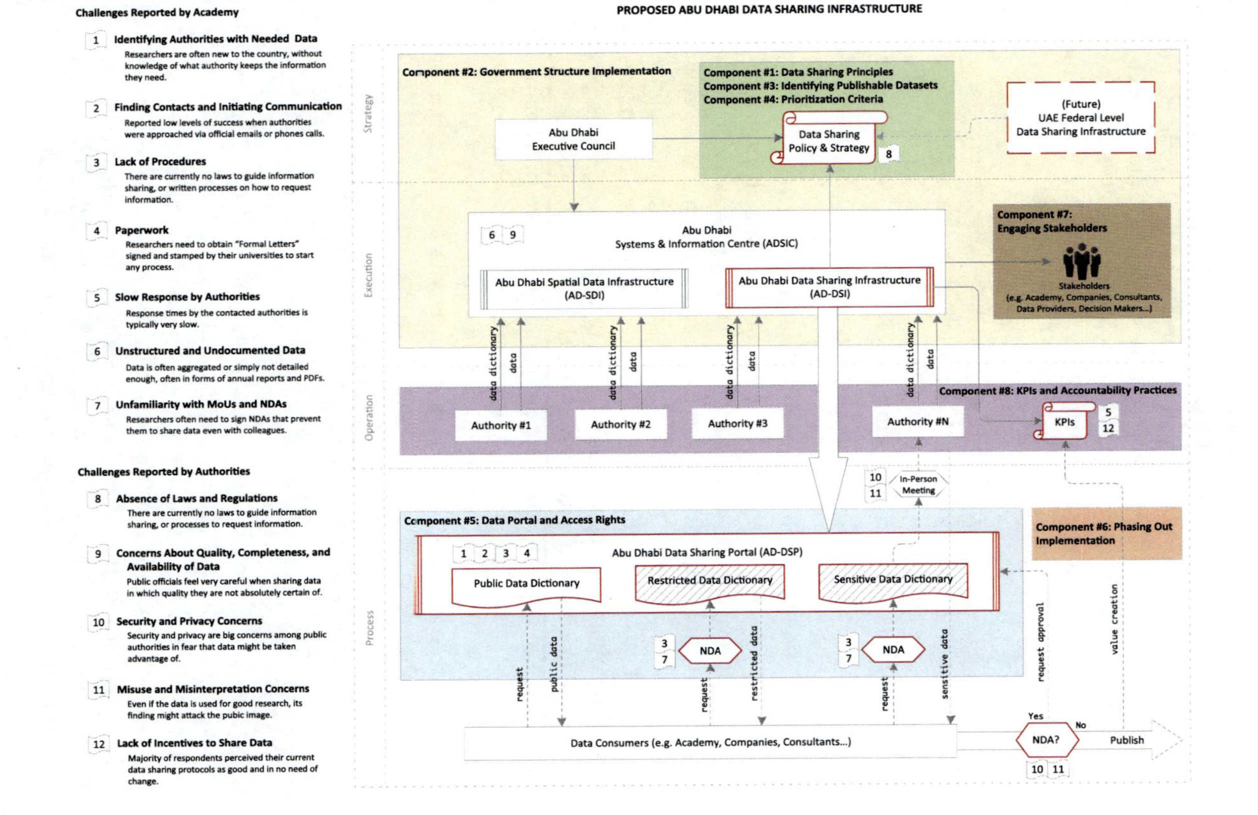

Fig. 2 Proposed solution framework

diagram, four layers of the implementation are marked with dashed lines to distinguish *Strategy*, *Execution*, *Operation*, and *Process* layers. White rectangles represent existing government entities or bodies. Outlined in red are newly proposed structures or protocol elements. Colored rectangles contain numbers to indicate how do the developed *Best Practices* and their respective adjusting recommendations fit in the big picture of the proposed framework (e.g., *Best Practice #1: Data Sharing Principles*).

4.3.1 Strategy

First and foremost, a policy directive at the level of the Emirate of Abu Dhabi needs to be initiated by the *Abu Dhabi Executive Council*. The *Executive Council* would, therefore, announce a *Data Sharing Policy & Strategy* to state the willingness of Abu Dhabi government to implement the new data-sharing infrastructure, as well as provide general guidelines for its execution. Following the arguments from *Best Practice #1*, such a policy should take a relaxed approach to introduce data sharing principles, focusing on improving access to already existing datasets, in their currently present forms, with permissions and usage rights that are already in place. A detailed strategy to fulfill those principles should follow, providing know-how for *Authorities* on how to identify which datasets are suitable for publication, as well as how to prioritize their release, as explained in *Best Practices #3* and *#4*, respectively.

In the future, the policies and strategies could also be responsibilities of a *UAE Federal Level Data Sharing Infrastructure*, if such happens to be developed. By introducing the *Data Sharing Policy & Strategy*, the following currently existing challenges are expected to be resolved or mitigated:

- Absence of Laws and Regulations

4.3.2 Execution

For the reasons explained in *Best Practice #2*, *Abu Dhabi Executive Council* shall appoint *ADSIC* to execute the implementation of the new data-sharing infrastructure. Using the know-how, signed agreements, and processes in place for their existing *AD-SDI* program, *ADSIC* shall create a new internal body—*Abu Dhabi Data Sharing Infrastructure (AD-DSI)*—to facilitate data sharing inside the Emirate. *AD-DSI* would be focused on fulfilling the principles described in *Best Practice #1* by further developing *Data Sharing Policy & Strategy* and coordinating its execution with public *Authorities.* Following the set *Strategy*, *AD-DSI* would standardize the common structuring and documenting practices, educating *Authorities* on how to prepare their datasets for faster and easier use by *Data Consumers.* In addition, *AD-DSI* would be responsible for creating and managing the *Abu Dhabi Data Sharing Portal (AD-DSP)*, also referred to as the *Portal.*

Once the system is operational, *AD-DSI* would start regularly receiving *data* and *data dictionaries* from public *Authorities*, managing them, and sorting for display on the *Portal.* At the same time, through the *Portal, AD-DSI* would receive data sharing *requests* from *Data Consumers.* Such requests would either be handled directly by *AD-DSI* if the data is available and all permissions acquired, or if not—routed to relevant *Authorities* for further processing. It is also recommended that *ADSIC* expands its *Stakeholder* engagement to people and entities involved in contribution to or usage of the *Portal.* This can provide *ADSIC* with insights and directions on how to best cater to its *Stakeholders'* needs and concerns, as further explained in *Best Practice #7.*

By placing a central body to coordinate *Authorities* in their data sharing and structuring practices, the following currently existing challenges are expected to be resolved or mitigated:

- Unstructured and Undocumented Data
- Concerns About Quality, Completeness, and Availability of Data

4.3.3 Operation

Local *Authorities* would, under the supervision of *AD-DSI*, execute the following main responsibilities: Produce and maintain *data dictionaries*—digital inventories that account for all data assets created or collected by the *Authority*; classify their datasets according to different access rights; respond to data sharing *requests* routed through *AD-DSP*, as well as proactively provide datasets to *AD-DSI*; and measure and report their progress based on introduced *KPIs.*

Producing *data dictionaries* would be the first step in the process towards a more structured and efficient approach to data sharing by each *Authority.* By having such inventories of all data assets already in possession, *Authorities* will have an overview of the amount, complexity, state, and form of available data to help them make a strategy for its further classification and publication.

A four-layer data classification approach is proposed based on *Best Practice #5*. *Public* data would be the data freely available to access and use by anyone for any purpose. It would be hosted on the *Portal* or listed for a release by request. *Restricted* data would be accessible only by *Data Consumers* whose host institutions have signed appropriate data sharing MoU with *ADSIC*. An individual might or might not need to sign an already prepared *NDA* before such data is used or processed as a publication. *Sensitive* data is data with very strict usage limitations that, on top of signing a specifically assembled *NDA*, require in-person meetings with the authority that owns it before a decision to disclose the requested information is made. *Confidential* data is not meant to be released for any reason and for any entity.

Finally, proposed in Table 3 are detailed examples of metrics suitable for tracking the performance of each *Authority*, assembled by using relevant metrics suggested by Hadjigeorge (2016) and Bonaguro

Table 3 Detailed example KPIs for measuring authorities' responses to data requests

Category	*Example KPI*
Quantity	Percent of inventoried datasets that are published
	Percent of inventoried datasets published by data classification type
	Percent of inventoried datasets published by priority level
	Percent of requested datasets published[a]
Quality	Percent of datasets updated with required metadata
	Percent of repeated requests due to the low quality of initially released data[a]
	Percent of repeated requests due to the misunderstanding of the initial request made*
Impact	Number of respondents indicating that open data has improved their analytical work
	Number of products or publications made with shared data
	Number of data requests by data classification type[a]
Timeliness	Response time satisfaction rate[a]
	Average time needed for initial response to each request[a]
	Average time needed for closing successful requests[a]
	Average time needed for closing unsuccessful requests[a]
	Number of meetings held with people that requested data[a]

[a]Newly proposed KPIs to track *Timeliness* and success rate of responses to data requests

(2016), together with additional indicators of *Timeliness* of responses, as suggested in *Best Practice #8*.

By having *Authorities* executing these newly proposed responsibilities, as well as introducing *KPIs* to hold them accountable for efficiently executing their data sharing duties, the following currently existing challenges are expected to be resolved or mitigated:

- Slow Response by Authorities
- Lack of Incentives to Share Data

4.3.4 Process

The *Process* layer of the proposed framework is oriented toward the functionalities of *AD-DSP* and its interaction with *Data Consumers*. Following the arguments from *Best Practice #5*, *AD-DSP* shall be mainly responsible for facilitating data exchange between stakeholders that need it, and authorities that own it. First, ***data dictionaries*** of ***Public***, ***Restricted***, and *Sensitive* data, as classified by each *Authority* (Table 2), would be listed on the *Portal*. *Public* data could be downloaded without any restrictions if already hosted, or requested for release if only the dictionary is provided. *Restricted* and *Sensitive data dictionaries* are, for security reasons, only available to the stakeholders that have signed appropriate MoUs with *ADSIC*. *Restricted* datasets can either be fully hosted on the *Portal* for direct download or have the requests routed to the authority responsible for a prompt reply. In both cases, a person requesting the dataset will be asked to sign an *NDA* before the dataset is disclosed. *Sensitive* data would never be hosted on the *Portal*, but only the ***data dictionary*** would be provided. For each ***request*** of such datasets, in addition to the need to sign an *NDA*, one or more *In-Person Meetings* with the *Authority* would be needed to explain the data usage needs and goals further. Such a procedure would closely resemble the current data release patterns. Finally, *Confidential* data would not be listed on the *Portal*, nor disclosed under any circumstance. To further counter any security, privacy, or misuse concerns, the mentioned *NDAs* could specifically require *Data Consumers* to ***request approvals*** for publishing any work based on previously disclosed datasets. Such ***approval request*** would be through the *Portal* routed to *AD-DSI* for final audits.

By listing ***data dictionaries*** of each *Authority* on *AD-DSP*, all relevant *Data Consumers* would be immediately aware of what types of datasets are available in which *Authority*, directly overcoming some of the main challenges of PSI access in the UAE, that are for the most part—communication challenges. Furthermore, in the cases when *Restricted* datasets are accessed, by clearly communicating access rights and *NDA* requirements, *Data Consumers* would save months of time and effort currently spent on dealing with paperwork, back-and-forth communication, and unnecessary in-person meetings. Only in the case of truly *Sensitive* data would the process fall back to the current procedures in place, preserving the option for *In-Person Meetings* when necessary, but making the protocol standardized and known to all participating parties. Finally, the products of any *Publication* derived from the shared data can be required to be submitted back to *AD-DSI* for use as an indicator of *value creation* in the *KPIs*. In summary, by introducing changes in the *Process* layer, the following currently existing challenges are expected to be resolved or mitigated:

- Identifying Authorities with Needed Data
- Finding Contacts and Initiating Communication with Authorities
- Lack of Procedures
- Paperwork
- Unfamiliarity with MoUs and NDAs
- Security and Privacy Concerns
- Misuse and Misinterpretation Concerns

4.3.5 *Implementation Timeline*

Following the arguments for a three-phase approach explained in *Best Practice #6*, the implementation timeline in Table 4 details how the development of the proposed solution framework could be executed to address the major issues at their earliest stage.

5 Discussion

Despite the announcement of a new Open Data law by the ruler of Dubai (Government of Dubai Media Office, 2015), fully complying with the international practices and principles remains challenging. Until such change comes to a realization, smaller steps are possible to advance toward that goal and help support AI research and applications. This study tried to reconcile the two worlds, proposing a unique implementation of Open

Table 4 Proposed action timeline for the first 36 months of AD-DSI strategy implementation

Month	*Action*
Month 1	*Abu Dhabi Executive Council* announces the *Data Sharing Policy & Strategy*
Month 2	*ADSIC* assembles the *AD-DSI* team
	AD-DSI Team develops strategy and guidelines for public *Authorities* to follow
	Each public *Authority* appoints one of their staff members as a data-sharing coordinator, presumably the person already in charge of communicating with *AD-SDI*, to execute new responsibilities
Month 4	*AD-DSI* team revises its strategy together with *Authorities*, documents future practices, communication channels, and *NDAs*
Month 6	Phase I Milestone—*AD-DSP* is announced. It contains the published guides, strategy, document practices, lists and descriptions of all public *Authorities* involved, and sets basic communication channels between data *Stakeholders*. The *Portal* contains no ***data*** nor ***data dictionaries***, but manually answers possible questions and routes data ***requests*** directly to ***Authorities***
Month 9	Each *Authority* creates a ***data dictionary***—indexed inventory of all data assets owned
	Together with *AD-DSI*, *Authorities* work on documenting datasets with metadata, classifying them by confidentiality and usage criteria
Month 12	Phase II Milestone—*AD-DSP* now contains ***data dictionaries*** of all datasets held and maintained by public *Authorities* in Abu Dhabi. The data is classified according to the proposed four-layer categorization criteria, with only ***Public***, ***Restricted***, and ***Sensitive data dictionaries*** displayed. No data is still online at this point
	Authorities start systematically responding to requests for data by *Data Consumers*
Month 18	*Authorities* start prioritizing datasets for future proactive release
Month 24	Phase III Milestone—*AD-DSP* is now a fully functional *Portal*, providing some ***Public*** datasets for immediate access while facilitating data exchange in cases where access is restricted
Month 27	*AD-DSI* continues working on improving the service, resolves communication bottlenecks by automating procedures, or employing more staff, addresses feedback received from *Stakeholders*
	Authorities observe demand patterns, start actively releasing ***Public*** and ***Restricted data*** directly to the *Portal* for instant future access
Month 36+	Future Milestones—*AD-DSP* starts improving year-after-year, following the increase of demand, received feedback, technological trends, and newest international practices

Data that expands into a more realistic semi-open platform. With smaller changes, such a platform could be similarly applied to other countries of the Arabian Gulf and broader Middle East. At the very least, this chapter hopes to inspire discussion on how innovation in the public sector can encourage better outcomes when best practices are not merely applied, but thoughtfully localized.

Since the proposed solution framework was designed specifically for the Emirate of Abu Dhabi, a possibility for future integration on a federal level should be further discussed. The framework currently depicts *UAE Federal Level Data Sharing Infrastructure* but does not describe its possible functionality in detail. One option for such a high-level entity is only to play an advisory role, coordinating the newly proposed policy on the federal level, but leaving the strategy execution fully to the appointed entities in each Emirate. However, such decentralization raises a practical question: How does one access datasets owned by the federal level authorities?

One approach suggests following the practice run by *ADSIC* with their current *AD-SDI* program for sharing geospatial data: Their board of entities already accommodates federal level authorities such as the Ministry of Energy or the Federal Authority for Nuclear Regulation, for example (Abu Dhabi Spatial Data, 2003). Staying compliant to this practice, the newly proposed *AD-DSI* and analogous appointed entities in the other Emirates could also accommodate for such federal level authorities. However, in the long run, the implementation of such a solution would carry possible risks of overlapping responsibilities and redundancy of datasets, which are best to be avoided.

Another approach would be to separate the functionalities of proposed frameworks by introducing another, federal level framework to accommodate only for information collected by the federal level authorities. This logical and functional separation would be easier to comprehend by *Data Consumers* but would possibly cause a larger constraint on the government's resources. As mentioned in the Literature Review chapter, a Linked Data portal providing references to datasets hosted in a decentralized fashion could serve as a compromising, but challenging solution. A decision on which approach to choose might even be guided by a definition of a federal Right to Information law if such is to be introduced in the near future.

Regardless of the chosen approach, the strengths of "vertical" integration with federal level authorities are many. Datasets unified and normalized across the seven emirates would unlock the information potential for an even greater benefit to the UAE society as a whole, creating a national-level data infrastructure that can accommodate various types of use cases. Furthermore, this "vertical" integration is expected to promote value creation at a "horizontal" emirate level. Correlation, patterns, and trends between activities of different emirates, but coming from unified sources, would be easier to detect and utilize, providing a one-stop shop for supporting research and innovation of national significance.

6 Limitations of Work and Future Research

There are several limitations of this study worth mentioning. First and foremost, the decision to perform data collection with qualitative interviews, while based on the authors' best judgment, certainly limited the study outcomes to a relatively low number of correspondents, as well as possibly caused a certain level of selection bias. Another limitation was the authors' decision to address the current challenges by following solutions specific to Open Data, while other approaches might be equally possible and potentially leading to different conclusions. The work might also be criticized for trying "too hard" in providing a compromise solution between some of the latest trends in government innovation and the currently ongoing traditional approach. Looking for such compromises might even stall the progress of the very innovation it is trying to promote.

These and other limitations are certainly expected to inspire future research on the relevant topics of data sharing practices in the UAE. This study focused on government-academia relationships and data sharing possibilities, while equally important are local government–industry and industry-academia relationships, currently unexplored. In the increasingly digital world, it would also not be surprising to see solutions combining data and knowledge across these sectors. Furthermore, additional strength to the arguments that inspired this study could be made by quantifying the damage done by the current status quo, revealing in detail how postponing its resolution hurts the local economy, innovation, and overall progress.

7 Conclusion

In the UAE, there is a strong challenge to accessing government data from within academia, which can be particularly limiting for research on data-driven fields, such as AI. Twenty-two academic researchers across ten UAE institutes were interviewed to explain their efforts to accessing needed information. Although the data requested was, according to respondents, not in any way sensitive, findings show a high level of dissatisfaction with the data acquisition process. From understanding responsibilities of many authorities in place, over contacting officials and setting up meetings, to handling administrative steps, researchers spend an unreasonable amount of time obtaining data they need for conducting experiments. Less than 50% of the participants reported to have obtained some data, but none was satisfied with quality, scope, or format of the received datasets—rendering them un-useful for scopes of their academic studies. Governmental policies in place are often making this process harder, while the existent ones promoting data sharing are not always put into practice.

To contrast these findings, 12 people across public authorities and an e-government consultancy company in the UAE were interviewed to explain the current government stance toward sharing PSI with third-parties, and in particular academia. Results indicate that although there are no public laws guiding data sharing practices in the country, certain protocols do exist and seem to be common among public authorities. When data is shared with the general public, it is typically aggregated and processed in the form of quarterly or yearly reports. As such, when more detailed datasets are needed for scientific research, government authorities should be approached formally. In that case, written consents by offices of both the requesting institute and the authority should be signed before further negotiation for data sharing release can follow. Researchers performing the study would then meet with responsible people within the authority to explain their study objectives and the details of specific data requested. After another round of approvals, the researchers are supposed to receive the requested information.

Current procedures, although straightforward, seem to be designed around concerns of data quality, security, and its potential misuse, rather than support of fast information access. The success of any data request is dependent on a series of factors such as signing MoUs, researchers' familiarity with the specific types of data collected, and the availability of staff within government authorities to help with data requests. Finally, the staff is expected to donate their time without proper incentives, while risking being accountable if the shared data is used improperly. Within this context, the idea of immediately accepting trends such as Open Data stands unrealistic to the interviewed participants. Moreover, PSI is still mainly perceived as property rather than the public good, which, in the opinion of the author of this study, is mostly the result of the lack of relevant data sharing policies. Such policies would ease the process for data requesters, protect confidential data, and protect the individuals responsible for handling such requests.

In the aim to propose solutions to the recognized challenges, the Open Data movement was recognized as a starting point in the thought process. Open Data policies and strategy documents from different sources were investigated to recognize patterns and best practices to its implementation. The study concluded that for the best possible outcome and benefit to the UAE society, international Open Data practices should not be simply copied, but altered to fit existing local procedures and culture that drives them. By looking at a typical Open Data strategy from within the local context, solutions were provided to address its possible implementation challenges in the UAE. Finally, a proposal of a comprehensive solution framework was assembled and revised together with different experts and public officials to provide recommendations for the successful implementation of data sharing practices in the country. The proposed solution framework focused on the specific setting of the Emirate of Abu Dhabi, but discussed its possible expansion and integration with the federal level by following vertically unified, separated, or a Link Data approach.

This study hopes to serve as a basis point for understanding how innovation within the public sector can focus on addressing the data sharing challenges identified in this study. Applying the proposed or similar recommendations would directly provide solutions to the limited accessibility of research data in the UAE—a critical step to truly support local academic efforts and strengthen the government-academia relations. Such actions would have the potential to exponentially improve national research outcomes, enhancing the UAE's aspiring transition to a knowledge-based economy that leverages the full potential of AI.

References

Abu Dhabi Department of Economic Development. (2009). *The importance of transferring into a knowledge-economy and anticipated social effect*. Retrieved from http://ded.abudhabi.ae/.

Abu Dhabi Spatial Data. (2003, October). *New entities on board*. Coordinates, 6. Retrieved from https://www.abudhabi.ae/.

Abu Dhabi Systems & Information Centre. (ADSIC). (2016). *Mission*. Retrieved from https://adsic.abudhabi.ae/.

AlAnazi, J. M., & Chatfield, A. (2012). *Sharing government-owned data with the public: A cross-country analysis of open data practice in the Middle East*.

Al-Hammadi, A. S., Al-Mualla, M. E., & Jones, R. C. (2010). *Transforming an economy through research and innovation*. University Research for Innovation, 185.

Ardichvili, A., Page, V., & Wentling, T. (2003). Motivation and barriers to participation in virtual knowledge-sharing communities of practice. *Journal of Knowledge Management, 7*(1), 64–77.

Arzberger, P., Schroeder, P., Beaulieu, A., Bowker, G., Casey, K., Laaksonen, L., et al. (2004). Promoting access to public research data for scientific, economic, and social development. *Data Science Journal, 3*, 135–152.

Attard, J., Orlandi, F., Scerri, S., & Auer, S. (2015). A systematic review of open government data initiatives. *Government Information Quarterly, 32*(4), 399–418.

Banisar, D. (2006). *Freedom of information around the world 2006: A global survey of access to government information laws*. Privacy International.

Beniston, M., Stoffel, M., Harding, R., Kernan, M., Ludwig, R., Moors, E., et al. (2012). Obstacles to data access for research related to climate and water: implications for science and EU policy-making. *Environmental Science & Policy, 17*, 41–48.

Bloomberg, M., & Merchant, R. N. (2012). *Open data policy and technical standards manual*. New York City Information Technology & Telecommunications.

Bonaguro, J. (2014). *Open data in San Francisco: Institutionalizing an initiative*. Retrieved from http://sfmayor.org/Modules/ShowDocument.aspx?documentID=425.

Bonaguro, J. (2016). *DataSF: Evaluation and performance plan for open data*. Retrieved from https://docs.google.com/document/d/1wvrSviKN8mYtxVVYCw7WohoujJjSFvSkY_Tj3ku8UMU.

Byat, A. B., & Sultan, O. (2014). The United Arab Emirates: Fostering a unique innovation ecosystem for a knowledge-based economy. *The Global Innovation Index* 2014, 101.

Castro, D., & Korte, T. (2015). *Open data in the G8: A review of progress on the open data charter*.

Cheong, L. K., & Chang, V. (2007). *The need for data governance: A case study.*
City of San Diego. (2014). *Open data policy—Implementation plan.* Retrieved from https://www.sandiego.gov/sites/default/files/legacy/pad/pdf/opendataimpplan.pdf.
Cohen, J., Dietrich, S., Pras, A., Zuck, L. D., & Hildebrand, M. (2014). *Ethics in data sharing.*
Commission of the European Communities. (1999). *Public sector information: A key resource for Europe.* Green Paper on Public Sector Information in the Information Society. COM (98), 585 final.
Conradie, P., & Choenni, S. (2014). On the barriers for local government releasing open data. *Government Information Quarterly, 31,* S10–S17.
Constant, D., kiesler, S., & sproull, L. (1994). What's mine is ours, or is it? A study of attitudes about information sharing. *Information Systems Research, 5*(4), 400–421.
Cress, U., Kimmerle, J., & Hesse, F. W. (2006). Information exchange with shared databases as a social dilemma the effect of metaknowledge, bonus systems, and costs. *Communication Research, 33*(5), 370–390.
Dawes, S. S., Vidiasova, L., & Parkhimovich, O. (2016). Planning and designing open government data programs: An ecosystem approach. *Government Information Quarterly, 33*(1), 15–27.
Department for Business Innovation & Skills. (2014). *Open data strategy 2014–2016.* Retrieved from https://www.gov.uk/government/uploads/system/uploads/attachment_data/file/330382/bis-14-946-open-data-strategy-2014-2016.pdf.
Drake, D. B., Steckler, N. A., & Koch, M. J. (2004). Information sharing in and across government agencies the role and influence of scientist, politician, and bureaucrat subcultures. *Social Science Computer Review, 22*(1), 67–84.
Eckartz, S. M., Hofman, W. J., & Van Veenstra, A. F. (2014). *A decision model for data sharing.* Electronic Government, 253–264.
Einav, L., & Levin, J. D. (2013). *The data revolution and economic analysis.* National Bureau of Economic Research.
Elbadawi, I. A. (2012). *The state of open government data in GCC countries.* 12th European Conference on eGovernment (ECEG 2012), Barcelona.
Emirates News Agency. (2019). *Abu Dhabi announces establishment of the Mohamed bin Zayed University of Artificial Intelligence.* Emirates News Agency, Abu Dhabi, UAE. https://wam.ae/en/details/1395302795116.
Fedorowicz, J., Gogan, J. L., & Culnan, M. J. (2010). Barriers to interorganizational information sharing in e-government: A stakeholder analysis. *The Information Society, 26*(5), 315–329.
Fioretti, M. (2012). Open data: Emerging trends, issues and best practices-a research project about openness of public data in EU local administration. In

G. Bottazi (Ed.), *Open data, open society*. Pisa: Laboratory of Economics and Management.

Garrison, B. (2000). Journalists' perceptions of online information-gathering problems. *Journalism & Mass Communication Quarterly, 77*(3), 500–514.

General Services Administration. (2012). *Open government plan*. Retrieved from http://www.gsa.gov/graphics/admin/GSAOpenGov20100407.pdf.

Global Open Data Index. (2015). *United Arab Emirates*. Retrieved from http://index.okfn.org/place/united-arab-emirates/.

Government of Dubai Media Office (2015). Law No. (26) of 2015 Regulating Data Dissemination and Exchange in the Emirate of Dubai. Government of Dubai, Dubai, UAE.

Group of Eight. (2015). *International open data charter*. Retrieved from http://opendatacharter.net/wp-content/uploads/2015/10/opendatacharter-charter_F.pdf.

Hadjigeorge, N. (2016). *Open data: Measuring what matters*. Retrieved from http://govex.jhu.edu/open-data-measuring-what-matters/.

Harris, R., & Browning, R. (2013). *Global monitoring: The challenges of access to data*. Routledge.

Harvey, F., & Tulloch, D. (2006). Local-government data sharing: Evaluating the foundations of spatial data infrastructures. *International Journal of Geographical Information Science, 20*(7), 743–768.

Headd, M. (2014). *Open data guide*. Retrieved from http://opendata.guide/.

Heeks, R. (2002). Citizen access and use of government data: Understanding the barriers. *Modern organizations in virtual communities*.

Huijboom, N., & Van der Broek, T. (2011). Open data: An international comparison of strategies. *European Journal of ePractice, 12*(1).

Janssen, K. (2011). The influence of the PSI directive on open government data: An overview of recent developments. *Government Information Quarterly, 28*(4), 446–456.

Kassen, M. (2013). A promising phenomenon of open data: A case study of the Chicago open data project. *Government Information Quarterly, 30*(4), 508–513.

Landsbergen Jr., D., & Wolken Jr., G. (2001). Realizing the promise: Government information systems and the fourth generation of information technology. *Public Administration Review*, 206–220.

Lane, J., Stodden, V., Bender, S., & Nissenbaum, H. (2014). The value of big data for urban science. *Privacy, big data, and the public good: Frameworks for engagement*. Cambridge University Press.

Lee, D., Cyganiak, R., & Decker, S. (2014). *Open data Ireland: Best practice handbook*. Insight Centre for Data Analytics, NUI.

Legal Aspects of Public Sector Information (LAPSI 2.0). (2014). *D2.1—Good practices collection on access to data*. PDF. Retrieved from http://www.lapsi-project.eu/sites/lapsi-project.eu/files/LAPSI_D2.1_GoodPracticesAccess(final).pdf.

Los Angeles Data Team. (2015). *LA open data policy and playbook*. Retrieved from https://datala.github.io/od-policy/.

Marks, P., Polak, P., McCoy, S., & Galletta, D. (2008). Sharing knowledge. *Communications of the ACM, 51*(2), 60–65.

Matheus, R., & Ribeiro, M. M. (2014). *Open data in the legislature: The case of São Paulo city council*. Open Data Research Network. Retrieved from http://www.opendataresearch.org/.

Montgomery County Government. (2014). *Open data implementation plan*. Retrieved from http://montgomerycountymd.gov/open/Resources/Files/OpenDataImplementationPlan_FY14.pdf.

Mopas, M. S., & Turnbull, S. (2011). Negotiating a way in A special collection of essays on accessing information and socio-legal research. *Canadian Journal of Law and Society, 26*(03), 585–590.

NYC OpenData. (2012). *NYC opendata technical standards manual*. Retrieved from https://cityofnewyork.github.io/opendatatsm/.

Open Knowledge Foundation. (2015). *Open definition*. Retrieved from http://opendefinition.org.

Open Knowledge Foundation. (2016). *Open data handbook*. Retrieved from http://opendatahandbook.org.

Prime Minister's Office. (2010). *UAE vision 2021*. Retrieved from http://www.vision2021.ae/.

Project Open Data. (2016). *Open data policy—Managing Information as an asset*. Retrieved from https://project-open-data.cio.gov.

Public Data Group. (2014). *Supporting the national information infrastructure*. Retrieved from https://www.gov.uk/government/uploads/system/uploads/attachment_data/file/329817/bis-14-969-public-data-group-open-data-statement-2014.pdf.

Seidman, I. (2013). *Interviewing as qualitative research: A guide for researchers in education and the social sciences*. Teachers college press. Chicago.

Sunlight Foundation. (2014). *Guidelines for open data policies*. Retrieved from https://s3.amazonaws.com/assets.sunlightfoundation.com/policy/Open%20Data%20Policy%20Guidelines/OpenDataGuidelines_v3.pdf.

Tauberer, J. (2007). *The 8 principles of open government data*. Retrieved from http://opengovdata.org/.

Thomas, D. R. (2006). A general inductive approach for analyzing qualitative evaluation data. *American Journal of Evaluation, 27*(2), 237–246.

Treasury Board of Canada Secretariat. (2014). *Canada's action plan on open government 2014–2016*. Retrieved from http://www.opengovpartnership.org.

Turner III, D. W. (2010). Qualitative interview design: A practical guide for novice investigators. *The Qualitative Report, 15*(3), 754–760.

UAE Government. (2017). *UAE strategy for artificial intelligence*. UAE Government, Abu Dhabi, UAE.

Ubaldi, B. (2013). *Open government data: Towards empirical analysis of open government data initiatives*. OECD Working Papers on Public Governance. No. 22, OECD Publishing.

Weill, P., & Ross, J. W. (2004). *IT governance: How top performers manage IT decision rights for superior results*. Harvard Business Press.

Weiss, R. S. (1994). Learning from strangers. *The art and method of qualitative interview studies*. New York.

Willem, A., & Buelens, M. (2007). Knowledge sharing in public sector organizations: The effect of organizational characteristics on interdepartmental knowledge sharing. *Journal of Public Administration Research and Theory, 17*(4), 581–606.

World Bank. (2016). *Open government data toolkit*. Retrieved from http://opendatatoolkit.worldbank.org/en/.

World Justice Project. (2015). *The WJP open government index™*. Retrieved from http://worldjusticeproject.org/sites/default/files/ogi_2015.pdf.

World Wide Web Foundation. (2015). *Open data barometer global report* (Second Edition). Retrieved from http://www.opendatabarometer.org/.

Yang, T., & Maxwell, T. A. (2011). Information-sharing in public organizations: A literature review of interpersonal, intra-organizational and inter-organizational success factors. *Government Information Quarterly, 28*(2), 164–175.

Zhang, J., & Dawes, S. S. (2006). Expectations and perceptions of benefits, barriers, and success in public sector knowledge networks. *Public Performance & Management Review, 29*(4), 433–466.

Zhang, J., Dawes, S. S., & Sarkis, J. (2005). Exploring stakeholders' expectations of the benefits and barriers of e-government knowledge sharing. *Journal of Enterprise Information Management, 18*(5), 548–567.

CHAPTER 4

Strategy for Artificial Intelligence in Bahrain: Challenges and Opportunities

Hesham Al-Ammal and Maan Aljawder

1 Introduction

According to a highly cited report from PwC published in 2017, "global GDP could be up to 14% higher in 2030 as a result of AI - the equivalent of an additional $15.7 trillion - making it the biggest commercial opportunity in today's fast-changing economy" (PwC 2017). Furthermore, these gains will mostly affect countries that have the capacity to utilize AI in their economy, as "the greatest gains from AI are likely to be in China (a boost of up to 26% GDP in 2030) and North America (potential 14% boost). The biggest sector gains will be in retail, financial services and healthcare as AI increases productivity, product quality and consumption" (PwC 2017). Such opportunities will also affect smaller economies within the world developing countries, especially for countries that will plan for building their capacity to utilize this technology.

Ever since Bahrain's independence, it has been attempting to limit its dependence on oil exports and diversify its economic portfolio. In the 1970s and 1980s, it initiated several large industrial projects such

H. Al-Ammal (✉) · M. Aljawder
University of Bahrain, Zallaq, Bahrain

E. Azar and A. N. Haddad (eds.), *Artificial Intelligence in the Gulf*,
https://doi.org/10.1007/978-981-16-0771-4_4

as ASRI (shipping) and ALBA (aluminum), as well as positioning itself as a major financial hub in the region (World Economic Forum 2019; UNDP and Derasat 2018; Cherif and Hasanov 2014). The country has recently been working on new initiatives that include digital transformation, strengthening entrepreneurship and attracting start-ups, hosting regional ICT services such as Amazon Web Services (AWS) center, among others. Within this context, forming a strategy to utilize AI to modernize the economy and boost both the public and private sectors is a major initiative for Bahrain. This vision has been boldly outlined by a recent Royal speech by HM the King of Bahrain. In the opening session of the legislative cycle on October 12, 2019 (Akhbar-Alkhaleej 2019), HM King Hamad bin Isa Al Khalifa directed the government to:

> create a comprehensive National Plan to ensure full readiness to deal with the digital economy's needs, by adopting and employing artificial intelligence technology in the productive and service sectors, through implementing the necessary systems and the technical infrastructure, and encouraging investments, to ensure full benefits of this within the national economy.

The Royal statement specifically indicated the need for "diversifying the national income sources according to Bahrain's Economic Vision 2030" (Akhbar-Alkhaleej 2019). This royal statement had an immediate impact on several sectors in Bahrain, including increased efforts by the Bahrain Economic Development Board, higher education, and public sector organizations. It bolstered the current efforts to leverage AI in the diversification of economic factors in Bahrain and will essentially be followed by efforts to realize this vision within all sectors in Bahrain. In this chapter, we present a possible roadmap toward consolidating all these efforts to form a national strategy toward harnessing the power of AI within the Bahraini economy.

2 A Brief Timeline of AI

Artificial Intelligence is not a new field, but has recently seen rapid and promising developments with the advent and spread of cloud, high-performance, and advances in machine learning (ML). The convergence of technology, science, and big data made it possible to utilize the power of AI. To provide the reader with some perspective of the history of the field, we present a brief summary in this section.

The ancient history of AI goes back to the early automata invented by the Muslim scholar ibn Al-Jazari, who in the early thirteenth century used innovative hydraulic switching, to build four musical robots that were programmable in one of the earliest demonstrations of robotics (Al-Jazari and Donald 1974). His writings include detailed diagrams and designs showing automata that are built for some defined purpose, as one figure was entitled "A Compendium on the Theory and Practice of the Mechanical Arts," dated 1315–1316 AD. It is reported that this implementation inspired early automata in the west, including work by Leonardo Da Vinci.

However, our modern understanding of AI dates more recently to the twentieth century, and Fig. 1 summarizes the history of AI, including its two substrates/branches: Symbolic AI and Statistical AI. The figure is an adaptation of (Ertel 2018) which was enhanced with more information and extended to cover the work done in the past decade. Some major new advanced applications of AI were also added to the top part of the timeline.

In the first half of the twentieth century, work on first-order logic by Gödel and Turing inspired the invention of abstract models of computers. This led to Turing's work on devising an intelligence test for automatic

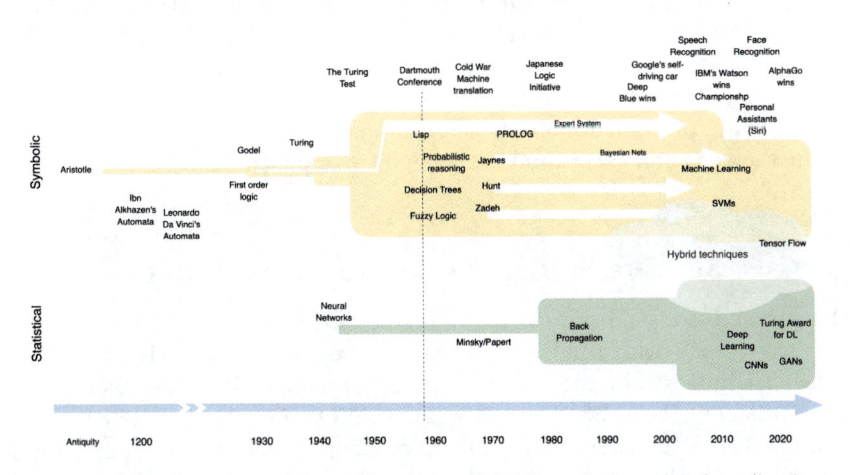

Fig. 1 A brief timeline of AI and its main subfields, techniques, and applications

computers in the 1950s (Ertel 2018). However, the first use of the term "artificial intelligence" was not coined until the popular conference at Dartmouth College in the summer of 1956. This conference then led to several advances within the field under the sub-field *Symbolic AI*. However, another branch was emerging in parallel, and this other main substrate of AI follows a different doctrine by simulating the working of the neurons in the human brain. In 1943, neurophysiologist Warren McCulloch and mathematician Walter Pitts used an electric circuit to simulate how neurons work (Minsky and Papert 1969). The advances in computers made it possible to simulate such neurons. Mathematically, these were modeled using statistical objects termed neural nets, which lead to *Statistical AI* with the following branches. This included techniques such as Neural Networks, Deep Learning, and Generative Adversarial Networks (GAN) (Goodfellow et al. 2014; LeCun et al. 2015).

It should be noted that since the late 1990s, *hybrid techniques*, which combine both the symbolic and the statistical worlds, were attempted to varying degrees of success. Recently, Tensor Flow (Abadi et al. 2016) hybrid techniques have been gaining popularity and success within machine learning. Another major observation regarding the success of the field of AI is that the road to the recent advances was not smooth. Many major failures of the symbolic branch lead to many researchers leaving the field. This occurred in the 1960s after the failure of machine translation, then in the 1980s after the failure of Japan's "Fifth Generation Project," among others. However, the statistical approach and its recent advances in putting forward real-life applications of machine learning brought the field back to prominence recently. The timeline in Fig. 1 attempts to highlight these pivotal events and the growth of applications within the field of AI.

3 The Singapore AI Strategy Model

Singapore has often been an economic role model for Bahrain, both being an island state with similar land size and population, and also being a financial hub within the region. Various observers, including members of Parliament (Yosif 2019), economists (UNDP and Derasat 2018), and journalists (Al-Obaidly 2008), often compare Bahrain to Singapore due to evident similarities that include the facts that both are a nation-state composed of an island archipelago (area around 750 sqm), that gained independence around the same time, and have limited resources. Furthermore, both island states have open societies and are considered as a top destination for expatriates. Cherif and Hasanov (2014) present the following insight into the economic development of both countries:

> Starting at comparable shares, the mining sector share increased in Bahrain from the 1990s relative to Singapore's. The manufacturing share in Bahrain is similar to that in Singapore, while the share of construction is slightly greater. In contrast, exports in Bahrain are almost exclusively concentrated in oil and metals (more than 95%), which is vastly different from the diversified export base of Singapore, which has more than 60% of total goods' exports in manufactures.

Following from this comparison between Bahrain and Singapore, we examine Singapore's AI ecosystem and its key players and drivers briefly. The ecosystem presented in Fig. 2 shows the key players in the AI industry and the main target sectors. It shows key components including research and development, ethical and regulatory frameworks, key players, and the current owner.

The key documents driving the introduction of AI to the digital economy in Singapore were produced in the 2017 Innovation Strategy, which was produced by Singapore's Committee for the Future Economy.

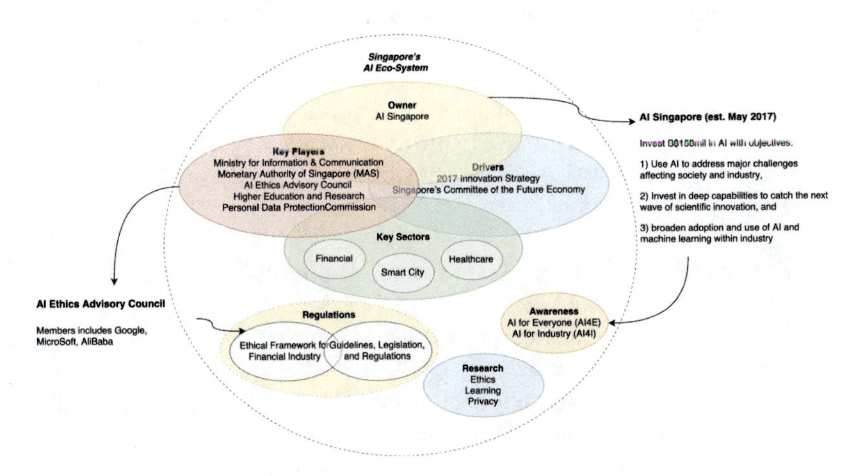

Fig. 2 Singapore's AI ecosystem

The Ministry for Communications and Information also named AI as one of the drivers toward growing the country's economy. The strategy identified three main sectors for AI: the financial sector, smart city management solutions, and healthcare.

In 2017, a national program called "AI Singapore" was initiated with a budget of S$150 million to harness the power of AI within the country. The AI Singapore program had several objectives which included: the use of AI to address major challenges in society, invest in catching the next scientific innovation wave, and broadening the adoption of AI and machine learning within the industry. In 2018, the program spawned two awareness programs: AI for Everyone (AI4E) and AI for Industry (AI4I), which aim at showcasing the capabilities of AI nationally (FLI 2019).

On the regulatory front, the Monetary Authority of Singapore initiated a consultation on the ethical pitfalls of AI in the financial sector at the end of 2017 (Lee 2017). This resulted in ethical guidelines for both data analytics and AI within Singapore.

Furthermore, in June 2018, an AI Ethics Advisory Council was announced, with 11 members including representatives from Google, Microsoft, and Alibaba. AI Singapore also announced a 5-year research program with the help of the Singapore Management University to assist the council in establishing ethical guidelines.

In addition, Singapore's Personal Data Protection Commission will establish principles "to ensure that decisions made with the help of AI can be explained, transparent and fair" (FLI 2019). Other research initiatives were also outlined by Singapore AI that span humanities, machine learning, and supportive disciplines such as privacy and security research.

Singapore's ecosystem for AI, which started in May 2017 and developed within the past two years, is illustrated in Fig. 2 and outlines the following key components:

- A single government program owner, responsible for implementing the strategy.
- Key players in the forming of the ecosystem.
- Drivers who steer the strategy.
- Key sectors being targeted for implementation of AI.
- Regulations and ethical frameworks.
- Awareness organizations.
- Research and development efforts.

The whole system is driven by collaboration and interaction among key government and research organizations affecting the main organizations within the public and private sectors. In addition, strategic planning and key frameworks initiated the subsequent work on AI implementation nationally.

4 Toward an AI Strategy for the Kingdom of Bahrain

This section investigates the current Information Technology industry in Bahrain by examining its main components. The aim is to identify key stakeholders for building an AI strategy for the future. Another aim is to identify key sectors for the initial introduction of AI techniques.

This scan will also attempt to discuss key strengths within the Bahraini IT ecosystem which can be adopted within the proposed AI strategy. This includes the recent efforts toward a healthy entrepreneurship ecosystem, especially within the tech industry.

Finally, as with the Singapore national plan toward harnessing the power of AI in Sect. 3, we present some of the key elements of a proposed AI strategy within the Kingdom of Bahrain, outlining key stakeholders, success areas, and possible sectors.

a. The Information Technology Ecosystem in Bahrain

Within the Kingdom of Bahrain, several entities drive IT and economic strategies. The major players include:

i. *Bahrain's Economic Development Board (EDB):* Ever since its establishment in 2000, the EDB has strived to implement a number of key initiatives, and it is the main national driver setting economic goals and policy implementing the Bahrain Vision 2030. The EDB promotes Bahrain for investment and supports diversification of economic activity. The current policy encourages, among others, start-ups culture and entrepreneurship (including the IT sector), finance and banking sector upgrade (e.g., FinTech), and a "business-friendly" Bahrain through the infrastructure and access to neighboring Gulf states markets. The Kingdom of Bahrain is promoted by the EDB as a major financial center in the region, including a vibrant Islamic Banking and Islamic Finance hub. The EDB initiatives lead to the liberalization of the telecommunications sector and the adoption of Bahrain as a regional AWS center for cloud computing.

ii. *The Information and eGovernment Authority (IGA):* Bahrain has one of the leading eGovernment entities in the region, and according to the United Nations in 2010, in the world. The IGA achieved government electronic integration and transformed government services in the past two decades. It also oversees the government data network, information, and national identity card. The IGA's two successive strategies (2007–2010 and 2012–2016) were behind the advanced state of electronic services and integration achieved by the government. The IT and services infrastructure created by the IGA contributes to the national digital transformation and to many facets of life in the Kingdom. Although both the 2007 and 2012 versions of the eGovernment Strategy did not include any reference to AI or AI technologies, the IGA is working on a new strategy which according to their leadership will include AI in the next version. Regarding the upcoming IGA strategy, the Authority's website states the following (The Information and eGovernment Authority 2019): "*The national strategies of Bahrain focus on advancing the living standard, taking into consideration the reduction of government expenses. The upcoming eGovernment strategy comes in line with this trend by ensuring online transformation in the government services through the usage of ICT which would facilitate knowledge management and completion of businesses professionally, conveniently, with low costs as well as ensure information security. The focus of the vision, message and objectives of the next strategy will revolve around creating a strong, flexible and safe environment to encourage innovation in public services.*"

iii. *Tamkeen:* Among the economic reform initiatives spearheaded by HH the Crown Prince, as part of Bahrain's Vision 2030, was the creation of the Labour Fund aka Tamkeen, which was established in August 2006. The public authority's role was to support Bahrain's private sector to position it as a key driver for economic growth. Tamkeen's two major objectives are: to foster and support the growth of enterprises and to enhance the training of the national workforce. As Tamkeen's 2018–2020 strategy focuses on the diversification of the training sector, it can be a major player in the capacity building efforts in an AI strategy. Its other role of fostering entrepreneurship can also serve the same purpose for AI start-ups, as well as the integration of AI within existing industries. In addition, Tamkeen provides several services to the private sector, including training financing, grants, entrepreneurship support, and counseling, to name a few.

iv. *Professional Societies:* Bahrain has several professional associations related to IT and AI, that are licensed under the Ministry of Labour or other regulators. They vary in terms of activity and membership size and include the following:

- *The Bahrain Information Technology Society (BITS)*: Founded in 1981, this is the oldest professional ICT society in Bahrain. Its main objective is awareness and training in ICT through conducting workshops, seminars, conferences, and lectures.
- *IEEE, Bahrain Section*: This is a local section from the Institute for Electrical and Electronics Engineers and is a professional society covering computing within its Computer Society. The Bahrain section is very active, especially among the University of Bahrain students majoring in Electrical Engineering and IT. It organized a Gulf-chapter level conference in Bahrain, with some tracks of research papers on AI. The Computer section is also becoming more active in recent years.
- *The Bahrain Society of Engineers*: The local professional organization for engineers with a membership of hundreds of active professionals in Bahrain established in 1971. The Society has an active membership and participates in several events regularly regarding the engineering profession.
- *The Artificial Intelligence Society*: A new society that was created in 2018 with activities ranging from awareness regarding AI to lectures discussing AI techniques. The Artificial Intelligence Society is described as "an independent and voluntary technological society formed in 2018 with the aim of promoting and disseminating Artificial Intelligence technology across the Kingdom." They currently have 40 members who include ICT company CEOs, AI professionals, and students (Artificial Intelligence Society Bahrain 2019).
- *The Technology and Business Association (TBS):* established in 2012 and works to build bridges between the ICT sector and

other high value-added sectors such as education and business management, focusing on the dissemination of ICT in those sectors and supporting the electronic development of national institutions to improve the level of electronic maturity in those sectors, and since its inception, the Society has sought to strengthen effective partnership with the private and public sectors. In October 2019, the association formed a permanent committee for Artificial Intelligence Technology.

- *The Bahraini Association for Researchers and Inventors (BRAINS)*: It was established in 2006 by the initiative of academic and administrative cadres, businessmen, and young people to empower researchers and inventors in the Kingdom by creating a supportive environment for scientific research and innovation and strengthening the communication network between the Society and other scientific research and innovation concerned bodies (Bahrain Brains 2020).

v. *The ICT Industry*: Bahrain is in the crossroad between east and west, has a vibrant educational system, and has always been a pioneer in ICT adoption within the GCC. Around a century and a half ago, the island had the first Indo-European telegraph cable installed in the MENA area back in 1864. It had many other firsts from that date in telecommunication and computing. Its pioneering work on the digital transformation of government services made it the winner of the UN survey in egovernment in 2010. Bahrain also boasts that it has one of the most liberal ICT markets in the region with the lowest cost of operating and living among GCC states. It offers foreign investors in the IT sector 100% ownership. It also has no restrictions on VoIP communications and has one of the best broadband services in the region. Current market emphasis is on: cloud computing, cybersecurity, digital content, e-commerce, and business services.

All of these factors lead to Bahrain being chosen as a hub for many international ICT companies, including Amazon Web Services (AWS), Huawei, Microsoft, Tata, and Cisco. There are also several large local ICT companies that partner with Microsoft, Cisco, and other vendors to provide solutions. None of the traditional big IT companies offered any AI-based solutions.

- *Batelco*: The national telecommunications company, which is partly owned by the government of Bahrain, also has several products and subsidiaries that offer IT solutions. Other than the telecom industry, it also provides solutions for healthcare, government, wholesale, and education, among others. The company is also part of FinTech bay founders and offers support for Bahraini start-ups.
- *01Systems*: Founded in 1986, this Bahraini software company was established by a pioneer local programmer in the early 1980s, Mr. Ali Sharif. It is one of the earliest software companies in the region and has developed several solutions for Arabization, banking, and document control. It offers many products for signature verification, biometrics, and document management. Currently, it provides several business management solutions in the financial sector and has over 300 customers worldwide. The company is one of the founding members of FinTech bay.
- *Kanoo Information Technology*: A family-based local company that has been in the Bahraini market for 20 years old. It partners with many traditional providers such as Microsoft, Trend Micro, and Dell, along with many security software providers. Most of the workforce are expatriates and have several contracts with public and private sector organizations.
- *Almoayyad Computers*: Another family-owned local company that has many international partners, including AWS, Microsoft, Avaya, Oracle, etc. The company has many awards and projects for mobile and integration solutions.

Recently, some ICT start-ups appeared in Bahrain. Some have been venturing into the AI market and are providing some innovative solutions. In the past 3 years, the compound annual growth rate (CAGR) for start-ups has grown to 46% in Bahrain, according to EDB sources. Furthermore, according to the EDB report: "digital start-ups, with 22% growth in Software and 13.3% growth in IT services are projected over the next 3 years." The EDB, along with Tamkeen, and other organizations have been fostering and pushing for a Bahraini ecosystem for start-ups. The following are some of the new ICT start-ups created within the past 4–5 years (Startup Bahrain 2019):

- *INFINITEWARE* is a local start-up specializing in AI solutions. Founded a decade ago, it started developing gaming solutions and gradually moved on in 2017 to artificial intelligence solutions. Its CEO, Mr. Ameen Altajer, describes it as an artificial intelligence company that provides products and services to several clients across the globe, INFINITEWARE has been working with large clients such as Saudi Aramco, STC (formerly VIVA), and the government of Bahrain. They also provide consulting services in other technologies such as automation, cybersecurity, and Blockchain.
- *CTM360* is a cybersecurity start-up headquartered in Bahrain, specializing in identifying and managing cyber blind spots outside an organization's network (surface, deep & dark web). They recently introduced BreachDB, which organizations can use as an online database that can help them view their organization's hacked credentials from over 100 + websites, 7.11 billion breached accounts, and 14 Million paste accounts picked up from the Deep Web.
- *Eat* was founded in 2014 and uses mobile computing to provide a restaurant booking experience.
- *WNNA* is a start-up that uses AI to provide contextual search engines through mobiles. It was founded in 2016.
- *Ingagrab* is another start-up company using AI and ML to help dealers navigate the international stock exchanges. It just started operations and in the early stages of being set up.
- *Malaeb* is a company in the early growth stage, a mobile app that provides a platform for renting sports facilities and courts.

The higher education sector is another stakeholder, and it has recently had some initiatives in strengthening the AI sector in Bahrain. This includes the establishment of postgraduate programs in AI at the University of Bahrain, as well as the Advanced AI Lab. The Bahrain Polytechnic also established an AI academy within its IT infrastructure.

The preceding key players in the IT ecosystem can act as the key stakeholders in any AI strategy for the Kingdom of Bahrain.

b. Bahrain's Tech Start-up Ecosystem

In an effort to scan the ecosystem being built in the past few years within the IT industry in Bahrain, Fig. 3 summarizes the major players

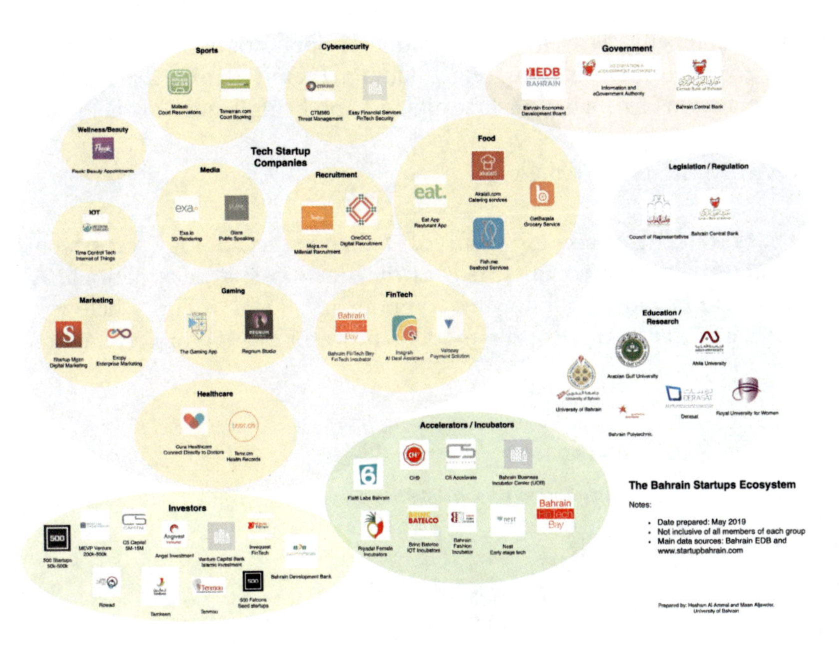

Fig. 3 The IT start-ups ecosystem in Bahrain in 2019

and resulting companies. The Tech Start-up Companies bubble includes the major start-ups categorized based on their specialization. The chart also shows the major investors, accelerators/incubators, educational institutions, legal and legislative organizations, and the main players from the government.

Although this currently shows an ecosystem for tech start-ups, it can easily leverage AI technology with the same components. Most of the current start-ups are within the IT field, which also supports the initiative to leverage AI within the economy. Another observation is that the majority are within the food/retail industry, especially after the success of the Talabat App in Kuwait and the GCC.

Overall, the number of accelerators/incubators is quite suitable for Bahrain, and among them, FinTech Bay is concentrating on financial technology, which is suitable for Bahrain's market with an emphasis on the banking and finance sector. From a quick analysis of the latest applications, there is a slight shift toward AI-related applications, especially for the banking sector. This is evident from the Ingrab AI dealer assistant start-up company, as well as a plethora of applications in the banking sector for customer relations (e.g., Bahrain Islamic Bank's Dana robot) or AI platforms for Bahrain's first digital bank being launched this year.

c. An AI Strategy Framework

Based on the preceding scan of Bahrain's technology ecosystem, including major components such as the start-up system, a national AI plan should include the following components shown in Fig. 4 below.

The main *components* of the model (shown in Fig. 4) for constructing the strategy are:

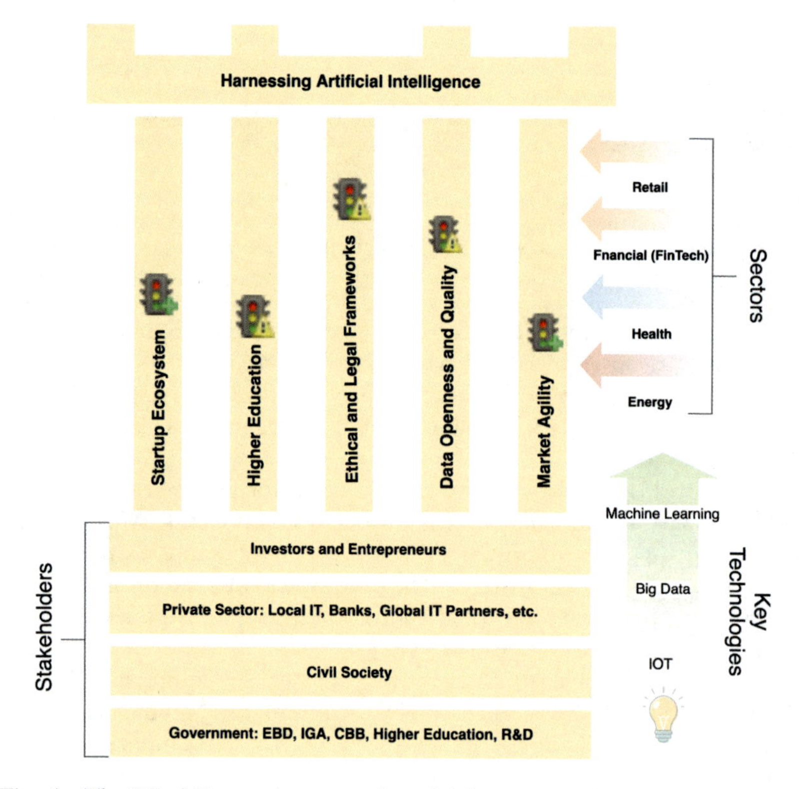

Fig. 4 The Wind Tower: A proposed model for constructing an AI strategy for the Kingdom of Bahrain

i. Stakeholders:

- *Government*: These include relevant organizations such as the Bahrain Economic Development Board (EDB), the Information and eGovernment Authority (IGA), the Central Bank of Bahrain (CBB), as well as major higher education providers such as the University of Bahrain and Bahrain Polytechnic. As discussed above in the current section, these government agencies can be the driving force behind AI in Bahrain through strategy leadership, pushing for regulation, as well as financing training through the labor fund.
- *Private Sector*: Both major local and international companies should be represented as stakeholders. Some of the mature start-ups should also be involved.
- *Investors and Entrepreneurs*: As a major catalyst for the ecosystem, investors should be involved in planning for an AI strategy. This includes accelerators operating in Bahrain, as well as venture capital organizations.
- *Civil Society*: Professional societies play a major role in securing talent and producing ethical frameworks.

ii. *Key Technologies:* Several technologies contribute to the success of AI techniques and thus need special attention within the strategy. These are all part of the digital transformation of society and include the following:

- *Internet of things (IoT)*: Sensing the world using IoT enables AI and machine learning to extend their reach to new domains. With the spread of IoT, this could boost the application of AI within the economy and is a major part of the digital transformation.
- *Big data*: "Data is the new oil" is a popular phrase nowadays. Most AI and machine learning algorithms depend on data, and thus the availability and quality of the data contribute directly to the success of AI.
- *Machine learning*: The recent success of AI was mainly due to the flourishing use of machine learning techniques. Higher education and IT curricula should pay attention to teaching ML techniques, including deep learning and GANs.

iii. *Target Sectors:* Among the sectors of the Bahraini economy, and from recent experience in the start-up ecosystem, it seems the following are the best candidates for implementing AI in the GCC market, and especially Bahrain: retail, financial, health, and energy.
iv. *Objectives/Pillars:* The main objective is, of course, harnessing the power of AI for the growth of the economy. The following are its main pillars.

- *The Start-up Ecosystem*: The fledgling start-up ecosystem in Bahrain can be used to leverage an AI strategy. Several companies can benefit from the boost of AI technology to expand their market reach and products.
- *Higher Education*: Universities and research institutes are the main producers of talent for an AI strategy. Plans for both skill and capacity building, as well as research and development, are a crucial part of a successful strategy. From observing the developments in academic programs in higher education in Bahrain, several programs have recently upgraded their curriculum with AI and machine learning courses. New postgraduate programs in big data have also been offered. However, there is still a big skills gap in AI, both in the professional and programmer level, and the academic staff level.
- *Ethical and Regulatory Frameworks*: A successful AI strategy cannot work in a vacuum. Especially with data and artificial intelligence, ethical and legal frameworks, as well as legislation to enable and protect users and producers, are vital. The strategy must include objectives and initiatives within this theme. Unfortunately, current regulations in Bahrain and the GCC do not cover most of the needed aspects of AI within society.
- *Access to Data*: This is a crucial pillar of any AI strategy. Under the supervised learning model, algorithms used in AI (especially in machine learning) use data to learn and to generalize. Without access to data, the algorithm is meaningless. Currently, massive amounts of data are generated by machines within the infrastructure. Unfortunately, within the GCC, this data is mostly collected by government ministries that are currently unwilling to use that data nor to share it. On the

other hand, data is also collected by private sector companies, and this is still largely not regulated within the GCC.

- *Market Agility*: Bahrain has a unique position to access several key markets within the GCC. This is especially true to the KSA market. The geographical and political ties with Saudi Arabia can enable companies operating in Bahrain to benefit from the market. Several major companies, such as Saudi Aramco and other gas and oil companies, may benefit from deploying AI applications.

5 Conclusion

Bahrain's position within the GCC, as well as its recent attempt to create a technology start-up ecosystem, can benefit greatly from a strategy toward harnessing the power of AI. The model presented above with key stakeholders, objectives, and sectors can be effectively leveraged to construct a national strategy for AI in the Kingdom of Bahrain. Some of the most important success factors include a mature eGovernment organization, a mature higher education system, an active Economic Development board, and a mixture of local and international IT companies. Other success factors include an open economy and labor force, with close markets in the GCC, and a young ecosystem for entrepreneurship.

From the environment scan conducted above, it is clear that AI technology has been largely neglected from IT strategy in Bahrain. It is also clear that without a national strategy that will unify efforts and target them behind the AI field, this will not be possible. The utilization of AI in Bahrain depends on a concerted effort among all stakeholders within the coming few years. IT start-ups and investors can be some of the biggest winners from such a strategy.

Several major pillars of any AI strategy need attention within the Bahraini society. These include the attitude toward data and its availability, especially from government organizations. More transparency is needed, as well as data stewards, open data, and regulations clarifying data ownership and privacy. This leads to another major aspect within the strategy, namely ethical and legal frameworks. The current active legislation regarding data, encryption, privacy, etc. is not sufficient for a modern digital economy. Although there is a law covering electronic crime, many of the required regulation is still missing or needs updating. This is true even for the banking sector, which is one of the most highly regulated sectors in the country.

After the recent Royal speech outlining a vision for utilizing AI for the Bahraini economy, a coordinated and strategic effort is needed by all sectors in Bahrain. This chapter outlines a possible model for such an effort that considers the current state of the economy as well as the major players within such a strategy. In order to validate this model, we organized interviews with key players within the public (eGovernment) and private sector (financial). The model mentioned above for building a national AI strategy was presented, along with a previous version of this chapter. The following points are a summary of the remarks by these decision-makers:

- Applications: Both public and private sector decision-makers indicated that one of the first applications they are investigating for AI is within "Customer Engagement." While the public sector seeks AI solutions that will boost government services and engage citizens seeking these services, the financial sector in Bahrain is looking for eBanking (eKYC or Know Your Customer). Several other AI applications were mentioned within the FinTech sector, such as aggregated data analysis, online loan eligibility and credit reporting, remote out-boarding, among others.
- Target Sectors: There was consensus on some of the proposed target sectors in the framework, namely: health and finance. There was a suggestion to rename the "retail sector" to "private sector" in the model. Also, the energy sector was supported to be a priority by some of the interviewees. There were suggestions to include the transportation sector as a candidate within the suggested framework.
- Stakeholders: There was agreement on the selected stakeholders within the framework. The interviewees emphasized the role of the IGA, EDB, and CBB. There were comments encouraging entrepreneurship and Bahrain's drive to be an attractive place for entrepreneurs and not just labor workers.

As a caretaker for an AI Strategy, there were suggestions at the national level for the Office of the First Deputy Prime Minister as the main owner of the process. Several suggested key stakeholders include the National Committee for ICT, Bahrain's EDB, and Bahrain's CBB.

- Strategic Objectives/Pillars: The interviews and discussions showed the need for a clear yet open Data Protection legislation that ensures privacy and other rights without stifling data use and openness. The government recently issued a law that has not been fully implemented yet and still needs the underlying procedures to be clarified. In terms of capacity building, there was consensus among the participants that there is a big gap in talent with the needed skills. This need must be met with local talent and transforming higher education, as well as attracting international ICT partners such as Microsoft, Google, and AWS. Clearly, applications in customer engagement have a head start with AI. Several banks have created solutions, including ABC Bank, which launched the first synthetic Digital Human using Soul Machines' Digital DNATM technology in September of 2019 (Bloomberg 2019). Bahrain Islamic Bank also announced a similar initiative.
- Technologies: There is a clear need for technologies such as voice and face recognition, cloud computing, big data, machine learning, and IoT for harnessing AI within the various applications in the private and public sectors. FinTech was identified as one of the major targets of AI technology, especially with the push by the EDB and CBB for growing this sector within Bahrain and its dependence on e-services.

Finally, grasping the opportunity of AI within any economy requires dedication and a strategic approach. As we saw earlier, Singapore's process included a single process owner, but also opened the floor for partnership nationally and internationally. Efforts within their process were serious, identified relevant sectors, and started producing regulatory and ethical frameworks. This effort must be adopted by one of the steering governmental bodies with partnership from the private sector, education sector, and civil society.

References

Abadi, Martín, Paul Barham, Jianmin Chen, Zhifeng Chen, Andy Davis, Jeffrey Dean, Matthieu Devin et al. "Tensorflow: A system for large-scale machine learning." In *12th USENIX symposium on operating systems design and implementation* (OSDI 16), pp. 265–283. 2016.

AI Singapore, Research, Retrieved May 2019 from https://www.aisingapore.org/research/.

Akhbar-Alkhaleej, Coverage of the Royal Statement in the Opening of the 2nd Legislative Cycle, Bahrain, 13 Oct. 2019. Available at: http://www.akhbar-alkhaleej.com/news/article/1186724.

Al-Jazari, Ibn al-Razzaz, and Donald R. Hill. *The book of knowledge of ingenious mechanical devices*. Reidel, 1974.

Al-Obaidly, O, "Between Bahrain and Singapore", *Alwasat Newspaper*, Issue 2291, 13 December 2008.

Artificial Intelligence Society Bahrain, Oil & Gas IQ, Retrieved May 2019 from https://www.oilandgasiq.com/events-crisisandriskmanagement/.

Bahrain Brains, Bahrain Brains Hackathon, Retrieved August 2020 from https://bhbrainshackathon.eventcreate.com/.

Bloomberg, Bank ABC launches the first synthetic Digital Human using Soul Machines' Digital DNATM technology, Bloomberg, Sept. 2019.

Cherif, R. and F. Hasanov, "Soaring of the gulf falcons: diversification in the GCC oil exporters in seven propositions". In Callen, M. T., Cherif, R., Hasanov, F., Hegazy, M. A., and Khandelwal, P. (2014). *Economic diversification in the GCC: Past, present, and future*. International Monetary Fund.

Ertel, Wolfgang. *Introduction to artificial intelligence*. Springer, 2018.

Goodfellow, Ian, Jean Pouget-Abadie, Mehdi Mirza, Bing Xu, David Warde-Farley, Sherjil Ozair, Aaron Courville, and Yoshua Bengio. "Generative adversarial nets." In *Advances in neural information processing systems*, pp. 2672–2680. 2014.

LeCun, Yann, Yoshua Bengio, and Geoffrey Hinton, "Deep learning," *Nature* 521, no. 7553 (2015): 436.

Lee, J., MAS steers debate on ethics of AI, Big Data; kicks off industry consult, Nov 28, 2017, Singapore. Retrieved May 2019 from https://www.businesstimes.com.sg/.

Minsky, Marvin, and Seymour A. Papert, *Perceptrons: An introduction to computational geometry*. MIT press, 1969.

Price Waterhouse Cooper (PwC), Sizing the prize: What's the real value of AI for your business and how can you capitalise?, June 2017.

Startup Bahrain, IT Ecosystem Companies, Retrieved May 2019 from https://map.startupbahrain.com/companies/.

Tamkeen, About Us, Retrieved May 2019 from https://www.tamkeen.bh/about-tamkeen.

The Future of Life Institute (FLI), AI Policy in Singapore, May 2019, Retrieved from https://futureoflife.org/ai-policy-singapore/.

The Information and eGovernment Authority, Our Strategy, Retrieved May 2019 from http://www.iga.gov.bh/en/category/our-strategy.

UNDP and Derasat (Bahrain Center for Strategic, International, and Energy Studies), National Report on Human Development, 2018. Available at http://hdr.undp.org/sites/default/files/bahrain_hdr_2018.pdf.

World Economic Forum, The Global Competitiveness Report 2019. Available at: http://www3.weforum.org/docs/WEF_TheGlobalCompetitivenessReport2019.pdf.

Yosif, Khalil, "On the Singaporean Experience", *Alayam Newspaper*, Issue 11027, 18 June 2019.

CHAPTER 5

Thoughts and Reflections on the Case of Qatar: Should Artificial Intelligence Be Regulated?

Ahmed Badran

1 Introduction

Technological advances can be regarded as a double-edged weapon. On the one hand, many befits can be reaped from the utilization of new technologies to improve the quality of life for human beings in different areas. On the other hand, the recent technological developments, particularly in the area of computing and robotics, raised a fundamental question about the possibility of the newly developed AI innovations to act independently from human control and to make their own decisions, which may harm humanity. In this context, different scholars and technology experts have echoed their concerns about the potential threats that AI may pose in the absence of government oversight and regulations (Reed, 2018). From an economic point of view, many economists share the fear that AI applications and machines alongside the advances in computing and robotics

A. Badran (✉)
Department of International Affairs, Qatar University, Doha, Qatar
e-mail: badranahmed@rocketmail.com

E. Azar and A. N. Haddad (eds.), *Artificial Intelligence in the Gulf*,
https://doi.org/10.1007/978-981-16-0771-4_5

may result in economic disruptions and higher rates of unemployment especially among low skilled workers (AI Forum of New Zealand, 2018). As such, AI applications are expected to result in job losses in all areas, including blue collars, white collars, and professional services (Russell & Norvig, 1994). At the same time, many activists and intellectuals are opposing the idea that governments should develop autonomous weapons and autonomous killing machines that work independently from human intervention and may select and destroy their own targets as this may result in significant security risks (Etzioni & Etzioni, 2017).

The ubiquity of AI in modern societies means that people, as well as governments, will be muddling through its legal and ethical ramifications for quite some time. The fast increase in AI applications raises fundamental questions about the potential impact of machines on the everyday lives of humans. As put by Scherer, 'with each passing month, AI gains footholds in new industries and becomes more enmeshed in our day-to-day lives, and that trend seems likely to continue for the foreseeable future' (Scherer, 2016). In this context, a valid inquiry would be whether AI applications will result in a better life and more efficient use of available resources, or they will pose threats, which might end humankind (Kohli, 2015). In general, AI cannot be seen as all good or all bad. It is a reality, it affects our lives in different shapes and forms, it provides opportunities, and it poses threats (Erdelyi & Goldsmith, 2018). The question now becomes, how could we deal with the AI threats in order to maximize the benefits and mitigate or minimize the risks? Addressing all the risks associated with AI goes beyond the scope of this chapter. Therefore, the chapter will focus on one aspect that is the policy and legal vacuum created by the AI revolution.

There are calls from scholars, AI practitioners, and technology leaders for a form of government regulation on AI activities and research in order to protect the public interest are gaining more attention. The founder of Tesler, Elon Musk, for instance, has regarded AI as being even more dangerous than nuclear weapons. In this regard, he wrote on Twitter, 'I'm increasingly inclined to think there should be some regulatory oversight [of AI], maybe at the national and international level'. In the same vein, regulatory and legal scholars, including Matthew Scherer, have called for the development of an overall legal and regulatory framework, which guarantees the safety of AI innovations through government intervention. The idea of developing policies and guidelines to regulate AI programs is not an alien even for the AI communities and industries. The Association for the Advancement of Acritical Intelligence has looked into this issue; however, the AI researchers have concluded that there is no need to develop such guidelines as the threats and risks associated with AI are not certain (Reed, 2018). Scherer has commented on the growing calls for regulating AI by stating that 'fear of technological change and calls for the government to regulate new technologies are not new phenomena. What is striking about AI, however, is that leaders of the tech industry are voicing many of the concerns' (Scherer, 2016).

Despite this growing agreement among several AI community members on the importance of government regulation and intervention, the question is still how much intervention is needed. In innovation and technology-driven sectors such as AI, too much government intervention and heavy-handed regulations might hamper the innovation and progress of these sectors (Finale & Kortz, 2017). Moreover, restrictive government regulations may result in less efficient AI systems, forced design choices, and suboptimal outcomes (Beishon, 2018). Hence, the regulation of AI will not be an easy task given the different meanings of AI in different areas and the risks the diverse forms of AI pose at different levels.

In this context, the chapter argues that recent developments in AI call for regulatory intervention from governments in order to strike a balance between potential benefits and the expected threats and risks. Nonetheless, any attempt to regulate AI is bound by the meaning we associate with this concept as AI means different things to different people and poses diverse types of risks in different policy domains. Moreover, the chapter emphasizes that we should not rush at present to restrictively regulate AI in ignorance. Instead, an incremental and gradual approach for regulating AI is needed wherein a distinction can be made between AI products and innovations that can be regulated within the existing legal and regulatory framework and those required new regulations. To follow up on this argument, the chapter will be divided into two main sections. Section one sets the stage for the discussion of AI regulation and the regulatory challenges posed by this novel construct. In this regard, AI and the other related concepts are discussed alongside the different positions taken on regulating AI from the leading AI entities. Section two is devoted to the discussion of a proposed regulatory framework to regulate AI in Qatar. The last chapter concludes with some policy recommendations on how to regulate AI without hampering innovation in such a promising and fast-growing sector.

2 The Complex Terrain of AI Regulation: Contextualizing the Debate

The growing body of literature and commentary on AI in recent years gives the impression that the issue of AI and its implications on societies is new. Despite the importance given to this issue by academics, innovators, technology leaders, and policy-makers, the concept of AI and the debate around its pros and cons go back at least a few decades. The term AI was firstly coined by John McCarthy in 1955 and the field of AI as a scientific discipline started a year later in 1956 (Calo, 2017). Additionally, the potential impact of robotics on employment has been a concern in the USA since the early 1980s when the New York Times made the headline 'A Robot Is after Your Job' (Shaiken, 1980). Some even may go further back regarding the debate on AI to refer to the controversies about the impact of computers and the new technological techniques in general on society. Regardless of how old the debate about AI is and the ways AI applications affect the societies, AI has become a reality and the debate about AI implications is attracting more attention from all stakeholders in AI policy communities. The main reasons for this are that, over the years, computational powers and access to data have dramatically increased and resulted in more opportunities for machine learning. Additionally, policy-makers are giving more attention to the issue of AI now in response to the calls coming from AI communities to develop the required policy frameworks and regulations (Calo, 2017). In this section, the main issues related to AI and its implications on societies will be discussed with a special focus on policy and regulatory concerns.

2.1 Defining AI

AI is a thought-provoking domain wherein science fiction is mixed with reality. What cannot be denied though, is the fact that we can see different manifestations of AI in diverse aspects of our daily lives from aircraft to virtual services, including nursing aids (Erdelyi & Goldsmith, 2018). AI applications have extended to cover many areas, and humans are becoming more and more reliant than ever on robots and other smart machines in manufacturing, healthcare, and even in finding answers to their most puzzling questions through electronic search engines. Despite such an acceleration of AI developments and applications in different domains, the concept itself lacks a clear definition. In other words, the idiom AI provides a wide umbrella, which covers several terms that are more specific (Stone, 2016).

In a general sense, artificial intelligence refers to 'a set of techniques aimed at approximating some aspect of human or animal cognition using machines' (Calo, 2017). In that sense, the cognitive element and the learning capacity of AI machines are what make them different from all other machines. Accordingly, AccessNow has defined AI as 'a machine with the ability to apply intelligence to any task, rather than a pre-defined set of tasks, and does not yet exist' (AccessNow, 2018). The crux of AI thus is to make machines more intelligent in the sense that they can act responsibly and appropriately, without harming the communities they operate in (Etzioni, 2018). From this perspective, AI can be conceived as 'an attempt to duplicate human intelligence, not to completely duplicate a human being' (Wang, 2008). In the words of Scherer, AI refers to 'machines that are capable of performing tasks that, if performed by a human, would be said to require intelligence' (Scherer, 2016).

AI can also be perceived in a narrow fashion, as 'the computerised analysis of data, typically very large data sets, to analyse, model, and predict some part of the world' (AccessNow, 2018). AI then has tangible technology, which includes software and hardware components (ILik, 2018). In that sense, AI refers to computers using data analytics in processing large amounts of data, in order to make decisions and come to conclusions, and most importantly learn from experiences (Etzioni & Etzioni, 2017). Consequently, many AI systems are programmed using a family of techniques referred to as machine learning (DATAx, 2019). In this context, many AI applications and machines, including Apple's Siri, Microsoft's Cortana, and Amazon's Alexa, are designed to learn from the behaviour of their users in order to serve them in a better fashion. Compared to old fashion AI notions such as 'symbolic systems' which looks at the machine's intelligence in terms of its ability to organize abstract symbols using logical rules, AI applications today are based on 'a particular set of techniques known collectively as machine learning' (Calo, 2017).

In this regard, machine learning can be perceived as 'the science of getting computers to learn and act like humans do, and improve their learning over time in an autonomous fashion, by feeding them data and information in the form of observations and real-world interactions' (Faggella, 2019). The machines' learning processes in this regard can include either supervised or unsupervised activities. In machines' supervised learning activities, a given algorithm is developed based on data, which are already labelled by humans. Conversely, in the unsupervised form of machine learning, AI software applications are freely left to find patterns without prior interventions by humans (AccessNow, 2018). The learning techniques themselves and the ideas of developing machines that can 'think' have always been around for decades (Russell & Norvig, 1994). Nonetheless, creating faster computers as well as developing better access techniques to big data sets have taken the learning capacity of AI machines to the next level (Stone, 2016).

Despite the advantages of AI and machine learning, one of the major concerns is the human ability to understand and control them. Many computer codes are presently readable and understandable by humans. Nonetheless, with entering the age of big data and analytics, and with the applied algorithms are getting much more complicated, the ability of humans to understand decisions and predictions made by AI machines is becoming far more restricted. As such, algorithms and the data sets behind them represent 'black boxes' for humans who attempt to unpack and understand their decision-making mechanisms (Mayer-Schönberger & Cukier, 2014). Different AI data analytics attempt to use complex algorithms in order to analyse big data and to underline and find patterns. Dealing with such high-volume, velocity, and high-variety information assets require cost-effective, innovative forms of information processing. As such, AI applications can enhance insight and decision-making (Gartner, 2019). The opacity of algorithms and the ways in which such black boxes operate are not a major concern from a regulatory point of view so long as they do not harm humans (Finale & Kortz, 2017). From this angle, the main goal of regulatory interventions should be to assure that AI systems behave in a way that is beneficial to people beyond reaching functional goals and addressing technical problems (Institute of Electrical and Electronics Engineers, 2016).

In the context of this chapter, AI will be perceived in accordance with Peter Stone's definition as 'a science and a set of computational technologies that are inspired by—but typically operate quite differently from—the ways people use their nervous systems and bodies to sense, learn, reason, and take action' (Stone, 2016). Compared to the aforementioned definitions, this one is broad enough to accommodate the different types of AI innovations. Additionally, this definition underlines the scientific nature governing AI developments and help to demarcate AI as a scientific field. With the meaning of AI so identified, the discussion moves now to focus on whether AI should be regulated or not, and how to control the associated public risks.

2.2 *Controlling AI Public Risks: To Regulate or not to Regulate? This is the Question*

To control the public risks associated with AI is to grapple with thorny ethical, social, legal, and regulatory issues. The problem with AI is that it has occurred in a regulatory vacuum. At the same time, there is a dearth of scholarship discussing potential regulatory approaches to AI (Scherer, 2016). The above discussion of the AI notions has indicated that a shift has taken place in contemporary AI applications and machines towards more focus on practical applications, which target very specific tasks relying on big data sets. Given the ethical, legal, and social consequences, the regulation of AI activities has become a necessity. In spite of the overall positive impact, people and policy-makers may hold regarding AI applications, the fundamental concern from a regulatory point of view is how technological advancements in this area be governed (Walsh, 2017).

In response to this question, Petit has made a distinction between two main approaches to AI regulation: *legalistic* and *technological* (Petit, 2017). From a legalistic perspective, the main concern is how existing laws apply to an AI application. The main areas that this approach examines include product safety and liability, cybersecurity, consumer protection, intellectual property, labour law, privacy, civil liability, criminal liability, legal personhood, insurance, and tax law. Unlike the legalistic approach, the technological approach to AI regulation does not take exiting laws and regulations as a given. Instead, it looks at the new AI innovations and sees if the issues they present call for the issuance of new regulations. A good example of this approach would be the Stanford Report investigating a one-hundred-year study on AI. The report has identified transportation, service robots, healthcare, education, low-resource communities, public safety and security, employment and workplace, home/service robots, and entertainment as the main areas to be impacted by AI applications. The report then specifies different legal and policy areas, which require regulatory attention. These areas include privacy, innovation policy, civil liability, criminal liability, agency, certification, labour, taxation, and politics (Stone, 2016). It is worth noting in this regard that these two regulatory approaches to AI are not mutually exclusive. That means AI regulators can use both approaches in dealing with the AI regulatory issues.

Under legalistic and/or technological regulatory approaches, AI activities can be governed by using different techniques, including industry self-regulation, ethics codes, contractual agreements, and government intervention (Brown & Marsden, 2013). Therefore, before talking about the rationale behind regulating AI, it might be a good idea to explore the different forms of regulatory options available. On the one hand, the practice in the field of AI has produced several initiatives under different names aiming at controlling and monitoring AI research and activities. Most of those initiatives are driven by the industry and carrying a voluntary nature. In other words, the AI industry now witnesses the emergence of a form of self-regulation based on some very basic notions of ethics. Case in point is the ethical codes in terms of 'a set of moral principles used to govern the conduct of a profession' (Collins English Dictionary, 2019). Such an unfolding development of professional ethics arising from the AI community is necessary but not sufficient for regulating such an important industry. Government regulation in the form of 'a specific set of rules backed by the sanctions of the state' is needed (Badran, 2019). In other words, ethical frameworks should be complementary to regulations (ILik, 2018). The reason for this is that ethics are normative and may have different meanings and contentions in different contexts. Additionally, the AI self-regulation means that the industry will be in charge of setting and monitoring the standards and rules controlling their own activities (Black, 2001). In such a case, there will be no guarantee that the public interest aspects are fully taken into account by the industry. Added to this, the enforcement mechanisms associated with AI self-regulatory initiatives, including code of ethics, are normally weak or absent in some cases. This lack of enforcement power encourages opportunistic behaviour from the AI industry, which may harm the interests of the people in the society (Calo, 2017). Put differently, the AI industry has the right to influence the regulatory policy-making processes using all the resources they may have at their disposal. Nevertheless, the task of forming and enforcing those regulatory policies should stay at the hands of the government in order to protect the interests of the public.

Talking about AI governance is another way to address the regulatory and policy concerns associated with the development and application of AI innovations. Applying the flexible notion of governance to AI may help to accommodate different structures, processes, actors, and modalities. It may also help to segregate the sectors that do not need as much intervention from governments such as technology from those, which require such interference. In other words, AI governance can help identifying the areas where we can govern AI innovations and activities, but without government (Rhodes, 1996). Despite the flexibility associated with the notion of governance, it seems too loose when compared with the precise notion of state regulation. In governance settings, and because of the increasing reliance on non-state actors and dispersion of powers among different types of stakeholders, it is difficult to identify what is being governed and by whom. Such lack of precision in a relatively newly emerging sector, such as AI, would confuse the picture rather than making it clearer.

2.3 *Regulating AI: The Rationale*

The literature on regulation and regulatory governance offers a plethora of justifications for state interventions in economic and social domains (Badran, 2012). For instance, state intervention and regulation can be justified on economic grounds in order to correct market failures (Breyer, 1982). From a social perspective, the state may use regulations to protect societal norms and values including human rights, freedom, privacy, justice, and many others in a way that helps build-up social trust (Brownsword, 2004). The state can also intervene through regulation in order to protect new industries and ensure their growth (Baldwin et al., 2012).

In the context of this chapter, the public interest argument for regulating AI sounds very plausible (Petit, 2017). Although it is difficult to clearly identify what is meant by the public interests as the interests of the public may take different forms in a different context (Badran, 2011). Moreover, acknowledging the fact that academics and practitioners may hold different opinions regarding the merits and value of state intervention, there is a public interest element in the area of AI that requires the intervention of the state in order to protect. The AI industry is dominated by big tech companies, which means the interests of newcomers to this market, or the existing small players might be in danger. We may disagree on whether governments are the best actor to represent the public interest in the area of AI. Nonetheless, without state regulation AI industry will not be willing to develop rules that benefit other parties at their expense. Furthermore, the lack of enforcement bodies and mechanisms in voluntary AI self-regulatory initiatives can weaken the development of the sector and encourage opportunistic behaviour from the industry. Considering the case of the developing countries, including the state of Qatar, regulation is a must, firstly, to secure the resources needed for the development of the AI industry, and secondly, to help building-up social trust in AI smart applications. Taken together, all these reasons justify a leading role for governments in regulating AI industries in collaboration with the rest of the stakeholders.

From a technological perspective, a strong rationale for regulating AI arises from the idea of *technological singularity*. Technological singularity can be defined as 'a point at which technological progress becomes unbounded' (Potapov, 2018). In that sense, intelligent computers and other forms of AI are reaching the point of being irrepressible. This technological singularity is expected to take place as early as 2030. At that point, AI will surpass human inelegance. Computers will become more capable than humans will and even far more intelligent. They will be able to produce new forms of technologies that outsmart humans and do not lend themselves to the traditional ways of control and regulation. In this regard, AI was seen as humankind's last invention (Barrat, 2013).

Considering that line of argumentation, it can be noticed that, it is true that AI computers and other machinery are becoming more intelligent and act smarter even than humans do. Nevertheless, AI machinery lacks a fundamental element that would prevent them from fully acting like humans that is motivation and desire. Unlike human beings who my try to pursue their desires and fulfil their motives once they are free of regulations and control, machines do not have desires which can direct them to act in certain ways. From this angle, the concerns about the domination of the machines over human sounds overstated. Added to this, when considering the regulation of AI, the pros and cons, including human and economic costs, must be clearly weighted-up.

Humanity is benefiting from AI applications in different areas including medical services and transportation. For instance, providing passengers' airplanes with AI programs prevent them from collisions and make them safer for travellers. The same can be mentioned about smart cars wherein the potentials of human errors have been greatly minimized.

In the area of healthcare, AI has helped in many ways even in operation theatres where robots have outperformed human surgeons in many tasks. As reported by DATAx 'As far as medicine is concerned, the bottom line is that systems able to use data effectively and learn from doing so are eventually going to affect every part of our lives – and perhaps even extend them' (DATAx, 2019). Given the case for and against AI, the question becomes: How can we take advantage of what AI systems have to offer, while also holding them accountable? (Finale & Kortz, 2017).

2.4 *The Characteristics of AI and the Potential Pitfalls of Government Regulation*

The characteristics of AI make it difficult for the government to intervene in the traditional reactive regulatory fashion in order to regulate its activities. The traditional regulatory methods including product licensing, research and development oversight, and tort liability will not be as effective and as productive in regulating AI developments, as is the case with other sectors. The first challenge for AI regulators is to determine what is meant by such a slippery concept (ILik, 2018). In other words, the first step to create common regulatory ground is to get some clarity on the notion of AI. This step is vital in order to know for a fact what to regulate in the field of AI. AI regulation is further complicated by the discreet, diffused, and opaque nature of AI activities, which render ex-ante government regulation ineffective. The autonomous nature of AI systems makes it difficult for governments to foresee and to deal with the unintended and unwanted consequences of AI developments. This is particularly true for certain AI systems such as autonomous weapons, which could result in human disasters if they act without control. In this context, the ex-post government regulation would also be unsuccessful (Scherer, 2016).

Despite the problematic features of AI, there is a good reason to believe that government regulation could be used to protect the public interest and to reduce public risks. Nonetheless, in such a complex and fragmented AI environment, the government intervention through regulation should be cautious and calculated in a way that avoids the possible negative consequences for the AI sectors, the regulators, and for end-users. Petit has identified some of the major pitfalls of government regulation for AI systems including disabling and knee-Jerk regulation, rent-seeking and regulatory capture, besides regulatory timing (Petit, 2017). Government regulation for AI can either be enabling or disabling for AI activities. Given the aforementioned benefits of AI applications, it would not be wise for a government to put in place too much regulation that stifles the development of the entire sector. A balanced approach is needed that weighs-up the pros and cons of government intervention in each case. Added to this, responding to potential regulatory risks by developing a heavy-handed regulatory system that prohibits specific AI activities or research in an automatic or unthinking fashion 'knee-Jerk regulation' would definitely impact negatively on the AI sector.

Given the financial ability and incentives of AI companies, rent-seeking and regulatory capture is also a possibility. The big tech companies in the AI sector have invested heavily in developing their systems and machines. This might push them to try to steer the government regulatory processes in order to serve their interests. The limited number of leading AI manufacturers and companies at this early stage of market development makes such a rent-seeking activity and regulatory capture a high possibility. Regulatory timing is another critical factor that may render government regulation ineffective. One of the major challenges facing the ICT regulators in general and the AI regulators, in particular, is the high speed at which technology develops. Regulators for decades are trying to catch up with technological advancements, but it seems they are still lagging behind. New technologies are developed at a faster pace than regulatory interventions, which makes government regulation always look like a fire fighting exercise. Hence, in the area of AI, 'the regulatory responses must be cognizant of that technological reality' (Brown & Marsden, 2013).

2.5 *AI Regulation and Government Intervention: The Current Position*

Despite the agreement on the need to address the social, economic, and ethical implications of AI, different governments around the world take different stands with regard to the regulation of AI. A common ground among all of them, though, is their agreement on the idea that it is too soon to regulate AI and that AI developments can be incorporated in existing regulatory frameworks. Hence, this section advances the discussion on the feasibility and drawbacks of government regulation of AI by looking at the different positions taken mainly by major players in the field of AI including the United States of America (USA), the United Kingdom (UK), and the European Union (EU) in an attempt to map out the regulatory approaches followed in this concern.

The US position can be summarized in the idea of promoting AI self-regulation with a limited role for government oversight. According to the report published by the National Science and Technology Council in 2016, the time is not right to issue an overall AI regulation, and it would be better if AI issues and concerns were included in the existing body of laws and regulations (The National Science and Technology Council, 2016). The way in which the US government has dealt with AI raised many concerns from the AI community. For many commentators and scholars, what the US government did is like fitting a pig into old square holes. The AI issues and concerns were treated as aliens and the US government tries to force them into existing laws and regulations (Cath et al., 2017). As such, the question of the ability of AI's self-regulatory rules to fit into the niche of existing regulations remains open.

Compared to the USA, the UK was keen on taking steps that are more drastic in dealing with AI-related regulatory issues. At the institutional level, the UK government has established new organizational bodies responsible for analysing and responding to the AI challenges. In this context, the AI Council was created as a platform wherein all AI community members, including industry leaders, developers, academics, and end-users, are represented. In this capacity, the AI Council acts as a podium for all stakeholders to echo and discuss their concerns about AI and to come to an agreement regarding the most pressing regulatory issues and the ways to deal with them. The UK government has also created the Centre for Data Ethics and Innovation to work as the government's advisory body on safety and ethical considerations related to AI. In the area of research and development, the Alan Turing Institute was instituted as the national research centre for AI. Additionally, a coordination mechanism was established under the name of the Government Office for AI in order to facilitate interaction and coordinate activities of all concerned AI bodies.

The discussions in the House of Lords' selected Committee on AI have indicated that the views regarding the timing and the shape of AI regulation have varied. On the one hand, some views have seen no reason for an AI-specific regulation at the time being as existing laws and regulations can accommodate the AI developments. On the other hand, some views have urged the government to intervene with immediate effect in order to develop and put in place a comprehensive regulatory framework for AI. The reason for this is that leaving AI with no AI-specific regulation would not help to address the unintended consequences for AI developments. In-between these two camps stand another group of people who proposed co-regulation as the way forward. The conclusions of the report were in favour of a government-led, gradual, and cautious approach for dealing with AI regulations. The idea of blanket AI-specific regulation was not very much welcomed, and an emphasis on the roles of relevant sectors' regulators has been underlined in relation to identifying the AI regulatory-related issues/needs for their sectors (The House of Lords: Select Committee on Artificial Intelligence, 2017).

At the EU level, there was an agreement among the member states on the importance of regulating AI in accordance with the core values of the EU. The stand of the EU on AI regulation is pretty close to the position taken by the UK government. The report published by the European Commission reinforces the idea that self-regulation could provide a good starting point; however; the report did not exclude the role of the EU commission entirely. The commission's role is to revise the existing legal and regulatory frameworks in order to make sure that the new AI developments are properly accommodated in a way that builds the public trust when dealing with AI applications (The European Commission, 2018).

2.6 *Regulating AI Systems: Stock-Taking*

Following on the above discussion, the question becomes: Is it really too soon to regulate AI? The answer is no. AI is already shaping our individual choices and affecting our lives in many shapes and forms. The ethical, economic, and social implications of AI developments can be seen in all domains. Having said that, it would be of great danger to leave such a vital rapidly growing sector without a proper regulatory framework. As a result, this chapter rejects the logic of the above-mentioned governments to deal with AI within the existing regulatory and legal frameworks. As put by Cath et al., AI 'is a powerful force that is reshaping our lives, our interactions, and our environments' (Cath et al., 2017). Hence, governments should not be treating AI the same way they treat other utility sectors and wait until it reaches the required level of maturity to intervene and regulate it. At that time, it might be too late for such an intervention and the intended regulatory outcomes may never be achieved. This problem is known as the Collingridge paradox (Petit, 2017).

Given the complexity and the fragmented nature of AI, the idea of leaving AI activities to sector regulators to handle might not be the best regulatory approach. As indicated earlier, AI provides an umbrella for different fields and scientific disciplines, including computer science, mathematics, information technology, and many other specializations in sciences and social sciences. In such a fragmented AI environment, a central regulatory mechanism is needed in order to secure harmonization and coordination. Hence, following a sector-specific regulatory approach in dealing with AI issues may result in more confusion in the sector as different regulators deal with AI issues in different ways, providing different and maybe conflicting solutions to the same AI problems.

From a public interest's perspective, there is too much at stake that requires government intervention in the form of AI regulations. AI applications may result in breaches in the areas of individual privacy and may expose personal data to unwanted third parties. Cybersecurity threats provide another reason for governments to intervene in order to minimize the risks of using AI applications. The users of AI applications might be more open to the idea of having a legitimate fair and impartial regulatory body that protect their interests rather than leaving the task of AI regulation to the industry itself because of the potential conflict of interests between AI developers and manufacturers on the one hand and the users on the other. To this end, AI regulators should enjoy certain powers chief among them is their ability to approve new AI applications and devices before entering the market. This will ensure the public that any devices or applications have been properly tested before being deployed to end-users. AI regulators should also have the power to sanction non-complaints and impose penalties and fines on AI stakeholders that do not conform to regulations. Such a balanced and centralized approach for AI regulation would secure the development of the sector in a harmonized and coordinated fashion while protecting the public interests by reducing the potential harm resulting from using AI applications and machines.

To sum up, what is needed is a regulatory regime that guarantees the safety of AI products and minimizes public risks. A regulatory regime encourages innovators and AI tech companies to incorporate built-in safeguards in all AI machines, which guarantee that the functions and the decisions made by AI machines such as driverless cars are conducted within a certain set of predetermined parameters. As such, the users will be assured that speed limits, for instance, will not be exceeded by autonomous decisions made by the car. The question now is who will guard the AI guardians? The answer is humans should be the guardians of the AI guardians. In order to avoid AI going too far beyond control, humans who are designing first layer and oversight AI systems must be in control. They should be able to shut down all AI systems if needed (Etzioni & Etzioni, 2017).

3 Regulating AI in Qatar

The Qatari population enjoys one of the highest rates in the world in using the internet (94%), and thus, have access to many AI applications (Qatar Center for Artificial Intelligence, 2018). Additionally, AI has been regarded as an enabler for other long-term economic, social, human, and environmental policy goals embedded in the Qatar National Vision (QNV) 2030 (General Secretariat for Development Planning, 2008). The analysis of the interview materials with experts in the AI community in Qatar has indicated that the AI community in Qatar is holding different positions regarding the need to regulate AI and the ways it should be regulated. These differences in opinions mirror the disagreement among the AI community worldwide regarding the issue under investigation in this chapter.

Based on the above, this section proposes some initial ideas on the way to regulate AI in Qatar. It is worth mentioning at the beginning that the discussion in this section does not aim at providing a well-developed and expounded regulatory framework for AI in Qatar. Instead, the aim here is to open the debate on the different regulatory approaches available for the Qatari policy-makers. The chapter suggests a regulatory mechanism encompasses two main components: an AI Law in addition to the creation of an Independent Regulatory Agency. This proposal is inspired by the previous successful regulatory experience in other related sectors such as Qatari telecoms and building on the ideas provided by Scherer (2016) in an attempt to put an initial stone in the wall of AI regulation in Qatar.

3.1 The AI Law

The absence of a designated law for AI in Qatar further complicates the debate about AI regulation. Hence, the first step in regulating AI in Qatar should be the issuance of an AI-specific law that lays the ground rules for such a fast-gowning sector and govern the behaviour of all involved actors. Similar to other laws in sectors such as telecoms, the AI law has to identify clearly the main players and the roles assigned to them. Chief among those players is the sectors' independent regulatory agency, which will be responsible for managing and controlling the public risks associated with AI innovations and applications. The AI law has to state the role of the agency and provide the AI sector regulator with the required legal mandate to manage the different regulatory aspects related to AI in Qatar.

The AI law also has to demarcate the boundaries between the regulatory agency and the other institutions in the field of AI in a way that allows the agency to act independently and to work for achieving the intended public interest policy goals. It is worth noting in this regard that the AI law represents the institutional guarantee for the regulatory independence of the agency. A strong and clear legal mandate should help the regulatory agency in exercising its powers in the face of all other institutions. It is also worth noting that the agency has an important role to play in translating the formal independence granted by the AI law to de-facto independence by working at arm's length from all AI stakeholders (Badran, 2017).

To compensate for the damage and the harm that might be caused by AI applications and devices, the AI law has to establish a fund designated for that purpose. The parties affected by AI faulty devices should have the right to file a case against the company that produced the faulty AI equipment. In case of accepting the claim, the affected parties have to be compensated for the harm from the fund. This fund could be financed from the citification fees paid by the AI tech companies for certifying their products. The fines imposed on the non-complying AI companies with the rules and regulations could be another source of finance for that fund. The fund could be managed by the regulatory agency in accordance with the AI law.

3.2 *The Independent Regulatory Agency for AI*

The mushrooming literature on regulation and regulatory governance provides a plethora of regulatory models and approaches available for policy and decision-makers. From those models, the chapter suggests the adoption of the Independent Regulatory Agency's (IRA) model wherein the legislators delegate their powers in rule-making and regulation to a specialized agency working within the remits assigned by the AI law. The debate about the rationale for delegating powers to independent agencies underlines several reasons ranging from political uncertainty at one end and creating credible policy commitments at the other (Badran, 2012).

Without going into the details of such a debate, the chapter suggests the argument for specialization is directly linked to the issue at stake. The crux of this argument is that governments delegate to IRAs because they are specialized agencies. That means those agencies are well equated in terms of resources and expertise for dealing with highly technical issues that other non-specialist parties could not handle as efficiently and as effectively as experts do. As such, the recruitment processes for the AI's IRA have to be designed in a way that attracts talented and qualified personnel with relevant AI educational and training backgrounds.

For the IRA to undertake its duties in full, the independence of the agency has to be institutionally secured (Badran, 2017). That means the agency has to be autonomous in making its own decisions without any form of unwanted interference from the industry. At the same time, the agency's decisions have to be insulated from any form of government intervention. By doing so, the IRA will be sending credible policy commitments to the AI industry that the decisions they make are founded on technical and public interest rationale rather than unjustified political inferences. Added to this, the independence of the regulatory agency-if fully exercised by the agency itself-will reduce the potential of regulatory capture and rent-seeking behaviour from the industry, as previously discussed.

Organizationally speaking, the IRA should have two main divisions: a board of directors and a technical certifying body. The board of directors will be responsible for setting the foundational rules for the overall agency's governance system. It also has to set the parameters within which the technical certifying division is working. The standards and rules for complying with public safety, the operational concept of public risks, and the notion of public interest have to be defined in a clear fashion by the agency's board. The members of the board should be appointed by the government taking into account their expertise in the field of AI or any other related fields. One of the first challenges the board as a policy-making body will face is to develop a working definition of AI for the purpose of regulation. As indicated earlier, AI means different things to different people working in different software and hardware fields. In such a context, the first step to regulate AI is to agree on what is being regulated. To this end, and given the importance of a working definition of AI for regulatory purposes, the agency may conduct a public consultation with AI stakeholders in order to come up with a definition that reflects the interests of all affected parties. Besides the board, the technical division will be responsible for setting the standards and technical criteria for citifying AI applications and devises in a way that minimizes the public risks and achieves the public interest in accordance with the mission and vision of the agency defined by the board.

With regard to the regulatory powers of the IRA, instead of having the power to prohibit certain innovations in the area of AI, the IRA should have the power to certify and license new AI applications and machines. This approach is less intrusive compared to other regulatory approaches, which may include heavy-handed AI regulations and in turn, may reflect negatively on the development of such an important sector. A certification mechanism to be developed and put in place for AI innovators should they wish to certify their products. The application processes and all required documents should be readily available on the agency's website. All AI developers wish to certify their products have to apply to the regulatory agency and support their applications with the evidence and test results showing that their AI innovations are risk-free and will not cause any harm to the public. Each application has to be assessed by the technical division of the agency, and the agency has to run its own tests in order to make sure that the AI products and applications will cause no harm to the public.

Following the assessment process by the agency's technical division, the regulator has the right to certify the product or to reject the application entirely. The agency should also have the right to ask for amendments to the original AI products before proceeding with the certification process. The agency could also certify certain AI products and devices to be used under certain conditions. For public safety, the agency has to make sure that all AI-related product-testing processes are done in an isolated environment. For new editions of previously certified products or for AI products similar to others, which have been previously certified by the agency, there might be a fast-track certification process. Following the certification of a certain AI product, the agency has to make sure via the licence agreement that none of the components of the certified AI product will be altered by the producing company or any other third party. By doing so, the agency will be sure that certified AI innovations are safe, and the public risk will be reduced.

In an effective AI regulatory system, different institutions play different roles. As a rule of thumb, the newly created regulatory agency needs to work in harmony with the other stakeholders and institutions, including the executive, the judiciary, and the legislature. Due to the lack of expertise, the legislatures are not well equipped to deal with and solve technical issues related to AI. Therefore, the legislative body normally delegates such rule-making power to the agency. In that sense, the agency can be perceived as a specialized policy-making body. This issue raises some concerns about the democratic legitimacy of decisions made by non-majoritarian institutions such as the IRA; however, these concerns are valid only in the context of western democracies and dealing with this issue here will go beyond the scope of this chapter. The judicial system plays a role as well in AI regulatory regimes via the adjudication of claims raised by individuals who suffer from the harm inflicted upon them by faulty AI devices and applications. The agency should also work with the executive branch of the government, but it should be working at arm's length in a way that does not jeopardize its independence. Finally, it also has to coordinate its efforts with other regulatory agencies in related fields such as telecom.

4 Conclusion

AI is one of the most continuous and consequential issues of our time. AI applications have far-reaching impacts and consequences on our lives that exceed what anyone could have imagined. As such, AI technologies have great potentials but also dangerous limitations. In this chapter, the issue of AI has been addressed from a regulatory point of view. The leading question was, should AI be regulated? If so, how? At the outset, there is no simple answer on whether AI should be regulated or not and how to regulate it. To answer this question, the chapter argued for stringent government regulation for AI activities and developments.

Despite current efforts led by the AI tech industry in relation to developing ethical codes, the discussion in this chapter has shown that this is an important but not a sufficient step towards regulating the AI sector. Most of those ethical codes have no legally binding basis. Added to this, ethical codes lack an enforcement mechanism via which rules can be enforced and violators can be sanctioned. These aspects make voluntary industry-driven codes weak and most likely lack the trust of the public. For these reasons, and because there is too much at stake when it comes to the way we should deal with the thorny issues posed by AI, a government regulation from our point of view is a necessity.

Government regulation for AI will help building-up the trust from the AI users as well as protecting the interest of the public and reducing the public risks associated with AI innovations. The chapter also emphasized the need for an innovative way to regulate such an innovation-led sector. In this context, instead of a prohibitive regulatory regime that includes direct intervention by the regulatory agency to prevent certain AI activities and products, the chapter proposed an incentive-based regulatory framework based on a certification mechanism.

It is worth noting in this regard that regulating AI is a highly challenging and problematic job. Many governments around the globe, including Qatar, are on the way to developing and putting in place different regulatory frameworks. These efforts are good as a starting point because policy-makers have to start somewhere. Nonetheless, the transboundary nature of AI problems does not lend itself to confined local approaches in dealing with such problems. Consequently, national efforts to develop AI policies and regulations should lay the foundations for establishing an overarching international regulatory framework. Such an international framework will guarantee the harmonization and coordination among the diverse national regulatory approaches.

In conclusion, AI has become a reality, and it manifests itself in different aspects of our daily lives. Now it is the turn of policy-makers to take the actions needed in order to control and regulate AI activities and applications. We need legally enforced regulations to be able to tussle with AI.

Acknowledgements This chapter was made possible by NPRP grant number 10-0212-170447 from the Qatar National Research Fund (a member of Qatar Foundation).

References

AccessNow. (2018). *Mapping Regulatory Proposals for Artificial Intelligence in Europe*. https://www.accessnow.org/cms/assets/uploads/2018/11/mapping_regulatory_proposals_for_AI_in_EU.pdf.

AI Forum of New Zealand. (2018). *The Potential Economic Impacts of AI Literature Review*. https://aiforum.org.nz/wp-content/uploads/2018/07/AIForum-DiscussionPaper-EconomicimpactofAILiteratureReview-May2018.pdf.

Badran, A. (2011). *The Regulatory Management of Privatised Public Utilities: A Network Perspective on the Regulatory Process in the Egyptian Telecommunications Market*. VDM Verlag Dr. Müller.

Badran, A. (2012). Steering the Regulatory State: The Rationale behind the Creation and Diffusion of Independent Regulatory Agencies in Liberalised Utility Sectors in the Developing Countries: Thoughts and Reflections on the Egyptian Case. *International Journal of Public Administration*, 204–213.

Badran, A. (2017). Revisiting Independence of Regulatory Agencies: Thoughts and Reflections from Egypt's Telecoms Sector. *Public Policy and Administration*, 66–84.

Badran, A. (2019). The Regulatory Policies of Telecommunication Sectors in the Arab Region: Selected Case Studies. *Siyasat Arabia (Arab Politics) Journal*.

Baldwin, R., Cave, M., & Lodge, M. (2012). *Understanding Regulation: Theory, Strategy, and Practice*. Oxford University Press.

Barrat, J. (2013). *Our Final Invention: Artificial Intelligence and the End of the Human Era*. Thomas Dunne Books.

Beishon, M. (2018). Is It Time to Regulate AI? *InterMEDIA*, 20–24.

Black, J. (2001). Decentring Regulation: Understanding the Role of Regulation and Self-Regulation in a Post-Regulatory World. *Current Legal Problems*, 103–129.

Breyer, S. (1982). *Regulation and Its Reform*. Harvard University Press.

Brown, I., & Marsden, C. (2013). *Regulating Code: Good Governance and Better Regulation in the Information Age*. Massachusetts Institute of Technology.

Brownsword, R. (2004). What the World Needs Now: Techno-Regulation, Human Rights and Human Dignity. In R. Brownsword (Ed.), *Global Governance and the Quest for Justice*. Hart Publishing.

Calo, R. (2017). *Artificial Intelligence Policy: A Primer and Roadmap*. https://lawreview.law.ucdavis.edu/issues/51/2/Symposium/51-2_Calo.pdf.

Cath, C., Wachter, S., Mittelstadt, B., Taddeo, M., & Floridi, L. (2017). *Artificial Intelligence and the 'Good Society': The US, EU, and UK Approach*. Springer Science Business Media Dordrecht.

Collins English Dictionary. (2019). https://www.collinsdictionary.com/dictionary/english/ethical-code.

DATAx . (2019). Guide To AI AND MACHINE LEARNING TRENDS IN 2019. file:///D:/Users/ab16495/Downloads/Ebook_____DATAx_Guide_To_AI_and_Machine_Learning_Trends_in_2019__1_.pdf.

Erdelyi, O., & Goldsmith, J. (2018). *Regulating Artificial Intelligence Proposal for a Global Solution*. Association for the Advancement of Artificial. https://www.aies-conference.com/wp-content/papers/main/AIES_2018_paper_13.pdf.

Etzioni , A., & Etzioni, O. (2017). *Should Artificial Intelligence Be Regulated? Issues in Science and Technology*.

Etzioni, O. (2018). Should AI Technology Be Regulated? Yes, and Here's How. *COMMUNICATIONS OF THE ACM*, 30–32.

Faggella, D. (2019). *What is Machine Learning?* https://emerj.com/ai-glossary-terms/what-is-machine-learning/.

Finale, D.-V., & Kortz, M. (2017). *Accountability of AI Under the Law: The Role of Explanation*. Berkman Klein Center Working Group on Explanation and the Law, Berkman Klein Center for Internet & Society working paper.

Gartner. (2019, 4 24). *Gartner IT Glossary*. Retrieved from https://www.gartner.com/it-glossary/big-data.

General Secretariat for Development Planning. (2008). *Qatar National Vision 2030*. Planning and Statistics Authority. https://www.psa.gov.qa/en/qnv1/Documents/QNV2030_English_v2.pdf.

ILik, S. (2018). *The Prospects for Successful Regulation in the Public Interest for Artificial Intelligence*. https://turkishlawblog.com/read/article/34/the-prospects-for-successful-regulation-in-the-public-interest-for-artificial-intelligence.

Institute of Electrical and Electronics Engineers. (2016). *Ethically Aligned Design: A Vision For Prioritizing Wellbeing With Artificial Intelligence And Autonomous*. The IEEE Global Initiative for Ethical Considerations in Artificial Intelligence and Autonomous Systems. https://standards.ieee.org/develop/indconn/ec/autonomous_systems.html.

Kohli, S. (2015). Bill Gates Joins Elon Musk and Stephen Hawking in Saying Artificial Intelligence is Scary. *QUARTZ*. https://qz.com/335768/bill-gates-joins-elon-musk-and-stephen-hawking-in-saying-artificial-intelligence-is-scary/.

Mayer-Schönberger, V., & Cukier, K. (2014). *Big Data: A Revolution That Will Transform How We Live, Work, and Think*. Eamon Dolan/Mariner Books.

Petit, N. (2017). *Law and Regulation of Artificial Intelligence and Robots: Conceptual Framework*. https://pdfs.semanticscholar.org/3f70/353ce297322a8029b95d8c734a7a6a95f749.pdf.

Potapov, A. (2018). *Technological Singularity: What Do We Really Know? Information*.

Qatar Center for Artificial Intelligence. (2018). *National Artificial Intelligence Strategy Ffr Qatar*. Qatar Center for Artificial Intelligence.

Reed, C. (2018). How Should We Regulate Artificial Intelligence? *Philosophical Transactions of the Royal Society A: Mathematical, Physical and Engineering Sciences*. https://doi.org/10.1098/rsta.2017.0360.

Rhodes, R. (1996). The New Governance: Governing Without Government. *Political Studies*, 652–667.

Russell, S., & Norvig, P. (1994). *Artificial Intelligence A Modern Approach*. Prentice Hall.

Scherer, M. (2016). Regulating Artificial Intelligence Systems: Risks, Challenges, Competencies, and Strategies. *Harvard Journal of Law & Technology*, 353–400.

Shaiken, H. (1980, September 3). A Robot is After Your Job: New Technology Isn't a Panacea. *New York Times*.

Stone, P. (2016). *Artificial Intelligence And Life In 2030: One Hundred Year Study On Artificial Intelligence*. https://ai100.stanford.edu/sites/g/files/sbiybj9861/f/ai_100_report_0831fnl.pdf.

The European Commission. (2018). *Artificial Intelligence for Europe*. the European Commission. https://ec.europa.eu/digital-single-market/en/news/communication-artificial-intelligence-europe.

The House of Lords: Select Committee on Artificial Intelligence. (2017). AI in the UK: Ready, Willing and Able? *the House of Lords*. https://publications.parliament.uk/pa/ld201719/ldselect/ldai/100/100.pdf.

The National Science and Technology Council. (2016). *Preparing For the Future of Artificial Intelligence*. https://obamawhitehouse.archives.gov/sites/default/files/whitehouse_files/microsites/ostp/NSTC/preparing_for_the_future_of_ai.pdf.

Walsh, T. (2017). *EU parliament: Consultation on Robotics and Artificial Intelligence*. https://thefutureofai.blogspot.com/2017/10/eu-parliament-consultation-on-robotics.html.

Wang, P. (2008). What Do You Mean by "AI"? *Frontiers in Artificial Intelligence and Applications*, 362–373.

PART III

Existing Opportunities and Sectoral Applications

CHAPTER 6

Knowledge, Attitude, and Perceptions of Financial Industry Employees Towards AI in the GCC Region

Muhammad Ashfaq and Usman Ayub

1 Introduction

Artificial Intelligence (AI) implementation in Gulf Cooperation Council (GCC) countries can boost the Middle East into new silicon hub and hydrocarbon-based economy. These countries are economically and financially strong with a collective nominal GDP nearing US$2trn in 2020 and average literacy rate around 94%. However, one cannot ignore the humanistic aspect of financial services industry practitioners in these countries. This chapter aims to analyse the knowledge, attitude and perceptions of professionals towards the use and implementation of AI in the Gulf Cooperation Council (GCC). According to Deloitte (2018), *AI refers to a broad field of science encompassing not only computer science but also psychology, philosophy, linguistics, and other areas. AI is concerned with*

M. Ashfaq (✉)
IU International University of Applied Sciences, Bad Honnef, Germany

U. Ayub
COMSATS University Islamabad, Islamabad, Pakistan

E. Azar and A. N. Haddad (eds.), *Artificial Intelligence in the Gulf*,
https://doi.org/10.1007/978-981-16-0771-4_6

getting computers to do tasks that would normally require human intelligence. Currently, the use of AI in the financial system is at its initial stages, but it is gaining momentum with the passage of time.

Nowadays, the efficiency of computer machines and their low operating costs is playing an important role in adopting AI by financial institutions (Shalf & Leland, 2015). The world has gone through the fourth industrial revolution and stepped into the fifth industrial revolution. AI and the Internet of Things (IoT) are considered as pillars in this revolution (Bogale et al., 2018). Sharma and Tiwari (2015) states, *The IoT is comprised of smart machines interacting and communicating with other machines, objects, environments, and infrastructures*. In other words, IoTs is making the world a global digital house where globally disbursed people are connected via internet define a community.

1.1 Application of Artificial Intelligence

The application of AI is multi-faceted and ubiquitous and thus has found application multiple fields in recent years, such as in healthcare, optical communication, network planning, logistics, financial industry, and media (Gunning & Aha, 2018). The core of AI's is machine learning that consists of algorithms used to process information as well as computing calculations and mathematical operations. The roots of AI systems lay in machine language, based on various complex algorithms that communicate with each other in high-speed based on given conditional logic. These technical procedures make it possible for computer machines to have super decision-making ability (Mata et al., 2018).

AI is expanding its horizon in the financial and banking industry, and in the view of Mannino et al. (2015) AI-based computer systems are *more efficient and superior than human experts* in processing information and computations. Thus, the usage of AI in the banking sector can make the operations more impactful and hassle-free (Manning, 2018).

The financial industry is one of the most critical sectors of the economy. AI has redefined the banking sector by the emergence of automation, blockchain, and FinTech (Manning, 2018). AI technology is integrated into mobile and internet banking, automated teller machines (ATM), cash deposit machines, SMS, and emails (Zawya, 2018). This integration provides an AI system with more data to analyse, allowing it to explore more information from the data.

In the current era of AI, its role is increasing because of the ability to value addition and efficient performance in the business process as its efficiency and reaction time is much better than a manual system to discover information from the database (Davenport et al., 2020). The essential areas that can use AI in financial services are portfolio management, customer services, data analysis, and risk management. Furthermore, analysing different economic scenarios to predict the future of the economic and financial industry to near accuracy is possible with AI.

1.2 *Artificial Intelligence in the GCC Region*

It is estimated that AI will contribute as much as $15.7 trillion to the global economy by 2030. The AI industry in UAE and Saudi Arabia would/could get nearly 14 and 12.4 per cent, respectively, of their GDP in 2030 taking the lead; both the countries are in the list of top 60 countries in the Global Innovation Index (GII)-an index measuring the innovation progress and performance of 127 economies. In 2018, Abu Dhabi unveiled a *Strategy for Artificial Intelligence* to implement AI in different sectors while Saudi Arabia plans to build Arabian Silicon Valley with an estimated cost of $500 billion for a futuristic megacity called *Neom* (Sophia, 2018).

Intel Corporation, IBM Corporation, Microsoft Corporation, Facebook, and Google LLC are a few big names contributing to AI in GCC. Moreover, AI is affecting the financial sector, as Gulf countries are among the wealthiest countries of the world, due to abundant reserves of oil and gas (Global Ethical Banking, 2018). The availability of financial resources and centralized government systems is attracting the international companies to provide IT services. This is catalysed by the stable economic conditions and relatively high penetration rate of information technology in the region (Augustine, 2018).

Various emerging technologies like data science, blockchain, and cloud computing are pursuing the financial industry not only to invest in these technologies but also to adopt these technologies for their usage and benefit. This development is transforming the financial industry in the Gulf region. The most significant impact of advancement in the world of AI is observed in the financial and banking sector (Manning, 2018). Therefore, AI will be transforming the banking industry permanently in profound ways for the coming years.

Chatbot is a software that automates the communication in a pre-programmed manner without human interaction. Chatbot is a source of value addition and allows for more efficient communication and problem resolution for customers. Its use is increasing with the passage of time as an application of AI. The computerized algorithm enables the Chatbot to formulate an appropriate response against any query from the customer (Manning, 2018). This 24/7 availability of services with more efficiency and less cost is alluring the international IT companies to increase their market share in the Gulf region (Sophia, 2018).

1.3 Use of Artificial Intelligence in Gulf Cooperation Council Countries

Governments, financial institutions, and businesses across the Middle East have realized ongoing technological disruption and the global shift of the traditional financial system towards sophisticated AI-based advanced financial system. Some financial institutions and banks have taken steps to adopt AI technology. These institutions are creating a confident atmosphere for other institutions to think about this technology and adapt it with the passage of time. According to Jain (2019), the Middle East is estimated to add 2% of the total worldwide benefits of AI in 2030. This benefit is equivalent to US$320 billion. Therefore, the penetration rate of AI is expected to be quite high in the future.

For instance, the Kuwait Finance House (KFH), the first Islamic bank established in 1977 in the State of Kuwait and one of the most notable Islamic financial institutions globally, has decided to implement AI Robotic Process Automation (RPA) to enhance business quality (IFN Fintech, 2018). In order to enhance the customer service experience, KFH has introduced AI Chabot developed by Microsoft technologies. This initiative will enable the bank to answer customer queries 24/7 and will support better decision-making by offering deep insights into the data collected (IFN Fintech, 2018).

Al Rayan Bank, a Qatar-based Islamic bank, is using cloud technology and is planning to incorporate AI-based management information and business intelligence system. In 2014, 85% of the bank's business was based on the latest information technology which was 57% in 2012 (Mckenna, 2015). Qatar is the most developing county of the Gulf region, and it is considered as the future of the Gulf region with a high possibility of leaving behind other countries of the region (Gremm et al., 2015).

One of the major hubs for AI penetration in the Gulf region is the Kingdom of Saudi Arabia (KSA). In absolute monetary terms, KSA is leading the region, and it is expected that KSA will have invested over US$135.2 billion in AI until 2030 that is equivalent to 12.4% of GDP. In relative terms of GDP, UAE is ahead of KSA, and it is expected to contribute 14% to the GDP in 2030 (Jain, 2019). While UAE is creating opportunities for AI adoption in its economy, this is expected to have a positive spillover effect on the overall economy by taking full advantage of available opportunities.

The Saudi Investment Bank is incorporating AI partnering SAP (a software company based in Germany) to facilitate its loyal customers (Argam, 2019). PayTabs, a Saudi-based payment solution provider, uses a combination of technologies, including AI to faster the payment processor such as remittances and SME transactions. PayTabs also aims to reduce financial fraud in payment processing. The scalability of PayTabs will affect financial institutions in the KSA to reduce the workforce and hence decrease the number of bank branches (Argam, 2019).

UAE-based Commercial Bank of Dubai is using AI to solve critical business problems, improve efficiency, and identifying new business opportunities. Since 2015, the RAKBANK is using AI to implement anti-money laundering (AML) solutions (George, 2018). Recently, the RAKBANK has expanded its 'strategic partnership to include the adoption of SAS Analytics' (SAS, 2016). Mashreq Bank, another prominent financial institution based in the UAE embracing AI technology in the area of consumer banking and it plans to reduce workforce up to ten per cent in the coming years (Mashreq Bank, 2018).

Alizz Islamic Bank of Muscat has incorporated AI system to manage their data. This system is known as Vision Banking Business Intelligence (BI) and is developed by Sunoida Solutions. Suleman Dossani, Managing Director and CEO of Sunoida Solutions, said *We have designed and built all important components based upon industry best practices within our Vision Banking BI solution, and we continue to innovate to provide the best services to our growing client base* (Alizz Islamic Bank, 2019).

One recent development in customer banking is from the Bank Muscat, which has engaged the AI system through SAS (a software technology company) to *implement customer relationship management (CRM) channel integration for their growing business demands'. The bank integrated multiple customer data touchpoints with the SAS Marketing Automation analytics engine to analyse the customer behaviour in cross-sell offers, digital offers, and promotions* (SAS, 2018).

One of the major factors that might impede the development of AI-based systems in GCC is the human aspect. These countries are characterized by a growing youth population, however, lacking in skills (El Shazly & Lou, 2020). Additionally, major hiring is carried out from SAARC countries (namely India, Pakistan, and Sri Lanka). This hiring is mostly of personnel with low computer literacy rate. World Bank projects that the GCC's labour force will exceed 20.5 million by 2020.

Therefore, this chapter focuses on the knowledge, attitude, and perceptions of customers towards AI use in the financial services industry in the GCC region. It takes into consideration the GCC countries such as the KSA, Kuwait, Oman, Bahrain, UAE, and Qatar. The research intends to analyse the future benefits and challenges for the financial services industry. This research will enable the stakeholders in the GCC region to ascertain knowledge, motivations, and behaviour of their employees to this new technology. The following are the main research questions:

1. What are the knowledge, attitude, and perceptions of financial services industry employees towards AI in the GCC region?
2. What will be the impact of AI on the economy in the future?
3. What are the prospects and challenges of AI?

The rest of the chapter is organized as follows: In Sections 2 and 3, the literature review is presented, hypotheses are discussed and research methodology is elaborated. Section 4 highlights results and discussion. Section 5 is the conclusion, policy implications and research limitations.

2 Literature Review

In the 1990s, AI came first to simply replicate and then improve upon human intelligence in pattern recognition and prediction. This was a different ambition which required breaking the shackles of traditional human resource (Shediac & Samman, 2010). Since the 2010s AI research has fast emerged as a dominant science passing milestones; as Sergey Brin, co-founder of Google says, *the new spring in artificial intelligence is the most significant development in computing in [his] lifetime.*

Likewise, use of AI in the financial and banking sector is attracting substantial attention. The availability of abundance of data by the passage of time and the future expectations for adopting AI is playing an important role to persuade entrepreneurs to invest in AI-based projects and ventures. The study by Eisazadeh et al. (2012) shows that in most countries in the Middle East public sector banks dominate the banking industry. The intervention of government in the private sector banking sometimes results in issues related to liquidity, credit, and interest rates.

A recent study conducted by the Alizz Islamic Bank in 2019 states that most managers focus on reporting aspects using a typical Business Intelligence (BI) system. However, to deliver successful business solutions in financial reporting, there are three main components to be considered; first, flexible and efficient extracting data from all sources, second is the banking data modelling, and the third is the dashboards and reports. The use of AI is not only helping the executive management to receive a current and accurate view of the business at any time through many interactive dashboards and Chatbot, but it also ensures that the compliance and regulatory issues are handled with utmost diligence and efficiency (Alizz Islamic Bank, 2019).

According to Verma (2017), AI can be essential in altering customer engagement with the banking sector in the Middle East. Some banks in the Middle East have deployed Chatbot for providing customers with a better-personalized experience. The use of Chatbots and other automation of systems in the financial sector with favourable outcomes are creating a conducive environment for other financial institutions to adopt this technology in the future (Verma, 2017).

Alzaidi (2018) studies the adoption of AI by employees. The study gathers the data from an extensive/widespread area, and a sample of 200 bank employees, collected from selected banks in the Middle East region. This study indicates that although the application of AI in the financial services industry especially 'banking sector is in quite an early phase, with the use of sophisticated algorithms', there could be risk that is more efficient and 'asset management in the banking sector that can further optimize financial policies of the Middle East banks'. Therefore, banks in the Gulf region can use fast and effective AI systems that can allow banking organizations to evolve more 'revenue generation models using smart financial AI management tools' (Alzaidi, 2018).

2.1 *Financial Market of Gulf Cooperation Council Countries*

The Gulf region has a relatively high per capita income due to an abundance of oil and gas resources. Shirish et al. (2016) state that this high per capita income propels cutting-edge information technologies, and there has been an increase in the demand for AI-based technologies. The study shows that there is a starting of a new era of digital banking in the Gulf region, especially in UAE, Qatar, and KSA. Millions of people of these countries are adopting AI-based technologies, especially in the financial and banking sectors. In banking and financial institutions, interconnected mobile banking systems by using wearables, tablets, and smartphones do most of the tasks from buying e-commerce products and services to make e-payments.

Shirish et al. (2016) cited the McKinsey research report and stated that UAE has a 92% internet penetration rate, and KSA has a 65% penetration rate. This study highlights that 80% of consumers favour the internet relying on tablets, personal computers, smartphones, and they visit branches and call service in urgent needs.

Although there is an optimistic view about AI in the Gulf region, however, cyber security is considered to be a significant threat/challenge for the financial institutions in this regard. The relationship between financial institutions and customers is based on trust, and any security concern can deteriorate this trust level. To address this issue, KPMG suggests financial institutions, especially banks, must consider this security issue from a customer point of view (Pera, 2018). Thereof, trust must be considered as an essential element to analyse the perception and future of AI in the Gulf region.

2.2 *Labour Market of Gulf Cooperation Council Countries*

The recent technological development has increased the urbanization in the world. According to UN forecasts, by 2050 the share of the urban population will reach 64.8%, while in high-income countries it will reach 88.4%. The GCC countries presently have relatively urbanized population; however, the UN forecasts that, by 2050, more than 90% of the population of these countries will live in cities. Moreover, these cities have potential to act as hub for digitalization of the economic. E-government development strategy is being devised to address this issue.[1]

GCC countries need to develop the necessary infrastructure to digitalize the economy, where human resource is an integral part. AI enables labour to become more productive in pretty much all of the activities and tasks that it performs. However, new technologies do not necessarily increase labour productivity. Automation technologies normally reduce the labour's share in value added, reduce overall labour demand because they displace workers from the tasks they were previously performing (Acemoglu & Restrepo, 2020).

AI creates jobs; however, technology also eliminates the traditional jobs creating an imbalance in workforce requirements. McKinsey reports that by 2025, the region reaches the same proportion of digital employment that the EU has today, then approximately 1.3 million new digital jobs could be created, including more than 700,000 in Saudi Arabia alone. McKinsey also reports that nearly 45% of the jobs will be technically automatable in 2019. Resultantly, AI could lay off 2.8 million full-time employees. At the same time, it could help save around $366.6 billion in wages.

Thus, a benefit is accompanied by a risk of disruption by new technologies. It is estimated that 37% of people have become 'technophobic' and fear their jobs could be taken over by robots and other AI software. This is more prevalent in expatriates working in GCC countries.

In general, there is a lack of research about the perceptions, attitude, and knowledge of professionals towards AI working in financial institutions. However, above-mentioned few studies highlight the current developments and importance of AI in the GCC region. Therefore, the present research aims to fill this research gap.

Based on the literature discussed above, following are the hypotheses of this study:

1. Financial services industry professional's knowledge of AI follows average of industry.

[1] UAE has UAE National Agenda 2021, Saudi Arabia has National Transformation Program 2030, Qatar National Vision 2030, Oman Vision 2040 as National Program for Enhancing Economic Diversification, Kuwait follows Kuwait National Development Plan 2035 and Bahrain has outlined Economic Vision 2030.

2. Financial sector experience level and knowledge of employees about AI is not affected by country, education level and financial sector in the GCC region.
3. The financial sector experience level of employees does not have a significant effect on the level of AI knowledge.
4. The financial sector experience level of employees and AI knowledge level does not have a significant effect on the level of future optimism of AI.

3 Summary of Research Methods

There are two methods used to collect data from professionals working in different financial organizations in the GCC. An online questionnaire was developed and it was circulated through professional networks in the GCC region. The same questionnaire in paper-form was also distributed using personal contacts at various institutions. The data was collected during April and May 2019. The questionnaire for this research consists of closed questions and mostly Likert scale-based questions. In total, 180 responses were obtained out of 280 distributed questionnaires, but 157 responses were taken in the analysis because 23 responses were incomplete. Data is analysed by using a statistical tool, namely SPSS. The data is collected from banks and insurance firms in the GCC region.

Analysis of Variance (ANOVA) shows a significant difference in the mean response of variables. It is used to check the significant difference among responses of respondents, obtained from the survey. The study used the Principal Component Analysis (PCA) technique to analyse the data. PCA technique enables us to obtain latent variables from the survey data, enabling the segregation of different variables into different groups based on appropriate theories.

The study used the PCA technique to reduce the data and constructed six variables to check the established hypotheses. The six constructed variables are; financial market work experience level, knowledge level about AI, future expectations from AI, transparency and efficiency in the financial system in the future due to the implementation of AI technology, impact of AI on the economy and security and ethical concerns related to AI in the future.

Cronbach's Alpha that illustrates about consistency among variables checks the consistency among different variables. The value of more than

0.7 is considered reliable for internal consistency. The study uses the Ordinary Least Square (OLS) technique to regress different variables.

4 Results and Discussion

This section provides information about various analyses conducted and additionally, interprets the findings. There are 23 questions in the questionnaire. Eighteen of these questions follow Likert scale, while five questions deal with the personal characteristics of the respondents.

Before discussing different inferential statistics, it is worth highlighting the descriptive statistics. There are 157 respondents in this study. Table 1 shows the descriptive statistics of the respondents such as professional background, education level, and country of origin. Table 2 shows information such as the level of experience of professionals, the number of years served in the financial industry and level of familiarity with AI.

Most of the respondents of the questionnaire were professionals having a bachelor's degree (64%), followed by a master's degree (34%). More than half of the respondents have less than five years of financial industry experience. With familiarity with AI, most of the professionals are familiar with the use of AI in the business industry (61%), followed by professionals who are familiar with both, the use of AI in business as well as with data science (25%).

Table 1 Descriptive Statistics of the Respondents' Professional Background

Service industry *Variable:*	*F*	*%*	*Education level* *Variable::*	*F*	*%*	*Country* *Variable::*	*F*	*%*
Advisory and consulting	5	3	A-Levels	1	1	Bahrain	30	19
Asset Management	14	9	Bachelor	101	64	Kuwait	22	14
Banking	81	52	Masters	53	34	Oman	22	14
Innovation Lab	2	1	Others	2	1	Qatar	21	13
Insurance	55	35				Saudi Arabia	30	19
						UAE	32	20
Total	157	100		157	100		157	100

Table 2 Descriptive statistics of respondents' experience and familiarity with AI

Financial industry service period *Variable:*	*F*	*%*	*Level of experience* *Variable:*	*F*	*%*	*Familiarity with AI and machine learning* *Variable:*	*F*	*%*
Less than 5 years	80	51	Poor	5	3	Not familiar	11	7
5 years	35	22	Below Average	30	19	From Data science perspective	11	7
10 years	30	19	Average	75	48	From a business and finance perspective	96	61
15 years	11	7	Above Average	37	24	Both	39	25
20 years	1	1	Excellent	10	6			
Total	157	100		157	100		157	100

The internal consistency of overall data was checked using Cronbach's Alpha. The Cronbach's Alpha for the data is 0.89, and that is more than 0.7, which is considered well enough for internal consistency. The study uses binary data to analyse the knowledge, attitude, and perceptions of respondents about AI. The respondents having above-average responses about knowledge of AI were assigned the binary signal of 1 and employees knowing below average are assigned 0. Table 3 shows the results of knowledge and familiarity with AI.

The probability is less than 5%, and it shows that the knowledge and perceptions of financial professionals about AI are above average, which means they have sufficient information about AI. It also shows that most of the employees of the financial sector consider that they have above-average knowledge regarding the use of AI in the financial sector.

Table 3 Knowledge and familiarity about AI

N	P	*Std. error*	*t-Statistic*	*Prob.*
157	0.62	0.04	3.03	0.01

The results of the test of equality to check the mean difference among different GCC countries and financial sector experience levels are shown in Table 4.

The results show in Table 4 that there is no significant difference in the mean level of experience among countries. The average financial industry level of experience does not depend on countries, but education and financial sectors have a significant effect on the level of experience perception. The results also show that employees having a master's degree consider that their financial industry experience and exposure to the use of AI in the financial industry is relatively higher than employees having a bachelor's degree. An employee having a higher degree is capable of obtaining more experience from professional work than the employee having a lower degree and education. Similarly, employees of the banking sector consider that their experience level is higher than employees of insurance and other firms. The reason would be that banking is

Table 4 ANOVA analysis of financial sector experience level

Countries	*N*	*Mean*	*Std. Deviation*	*Std. Error Mean*
BAHRAIN	30	0.107268	0.855942	0.156273
KUWAIT	22	−0.237362	0.980303	0.209001
OMAN	22	−0.046693	0.693141	0.147778
QATAR	21	0.129264	1.207337	0.263463
SAUDI ARABIA	30	0.12293	0.824243	0.150486
		F-test		
Value: 0.648810			Probability: 0.6628	
Education	*N*	*Mean*	*Std. Deviation*	*Std. Error Mean*
Master level	53	0.273165	0.988284	0.135751
Bachelor level	101	−0.173757	0.843626	0.083944
		F-test		
Value: 8.653023			Probability: 0.0038	
**Education*	*N*	*Mean*	*Std. Deviation*	*Std. Error Mean*
BANK	81	0.138326	1.013418	0.112602
INSURANCE	55	−0.296545	0.740393	0.099835
		F-test		
Value: 7.427384		Probability: 0.0073		

*Two categories of education were ignored due to lack of observations

widespread with its scope being more comprehensive compared to insurance. The ANOVA results for the level of knowledge of AI are shown in Table 5.

The results in Table 5 show that the level of knowledge about AI does not depend on the country, education or in the financial sector. The reason would be that as use of AI in the financial sector is at its initial level, and the personal interest of employees is the primary factor to be familiar with AI and usage of AI in different sectors.

The relationship between the level of financial sector experience and AI knowledge is shown in Table 6.

The result shows that there is no significant relationship between the level of work experience and AI knowledge. There is a possibility that an employee having lesser experience in the financial industry may have more knowledge than the employee with more experience. This finding is

Table 5 ANOVA analysis of knowledge of AI

Countries	*N*	*Mean*	*Std. Deviation*	*Std. Error Mean*
BAHRAIN	30	−0.046979	1.028304	0.187742
KUWAIT	22	−0.385382	0.988194	0.210684
OMAN	22	−0.070261	0.689699	0.147044
QATAR	21	0.316569	0.847389	0.184915
SAUDI ARABIA	30	−0.017904	0.748515	0.136660
UAE	32	0.166334	0.714600	0.126325
		F-test		
Value: 1.807979			Probability: 0.1145	
Education	*N*	*Mean*	*Std. Deviation*	*Std. Error Mean*
Bachelor	101	0.011975	0.812991	0.080896
Master	53	−0.014206	0.943652	0.129621
		F-test		
Value: 0.032221			Probability: 0.8578	
**Education*	*N*	*Mean*	*Std. Deviation*	*Std. Error Mean*
BANK	81	0.086482	0.792958	0.088106
INSURANCE	55	−0.043068	0.857218	0.115587
		F-test		
Value: 0.818715		Probability: 0.3672		

*Two categories of education were ignored due to lack of observations

Table 6 Regression analysis

Variable	*Coefficient*	*Std. Error*	*t-Statistic*	*Prob.*
Constant (intercept of OLS regression)	−6.73E-18	0.06839	−9.84E-17	1
Financial sector experience level	0.040918	0.074374	0.550161	0.583

in support of previous results that knowledge about the use of AI in the financial industry is mostly due to self-interest, and it does not depend on education.

The relationship of the financial sector working experience level with other constructed future expectations of AI variables is shown in Table 7.

The results manifest that there are significant relationships among the variables with the financial sector work experience level mentioned in Table 7. It means that if an employee possesses more or additional financial sector knowledge, the more he or she is optimistic about the future of AI in the financial sector, positive impact of AI on the economy and work efficiency and transparency in the system due to the incorporation of AI. Meanwhile, the employee is more concerned about ethical and security issues related to AI in the future.

The results in Table 8 illustrate that there are no significant relationships of all the above variables with the AI knowledge level. It means that

Table 7 Dependent variables on financial sector experience level results

Dependent variables	*Coefficient*	*Std. Error*	*t-Statistic*	*Prob.*
Future expectation	0.396946	0.059946	6.621765	0.0000
Positive impact on economy	0.321104	0.062572	5.131752	0.0000
Work efficiency and transparency	0.429019	0.078796	5.444700	0.0000
Ethical and security issues	0.224078	0.061478	3.644820	0.0004

Table 8 Dependent variables on knowledge of AI results

Dependent variables	*Coefficient*	*Std. Error*	*t-Statistic*	*Prob.*
Future expectation	0.051445	0.073139	0.703389	0.4829
Positive impact on economy	0.115339	0.072430	1.592407	0.1133
Work efficiency and transparency	0.072257	0.092607	0.780249	0.4364
Ethical issues	−0.015331	0.069103	−0.221855	0.8247

employees knowing AI are indifferent about the future of the use of AI in the financial sector, the positive impact of AI on the economy, work efficiency and transparency in the system due to AI and ethical and security issues related to AI. A summary of the constructed hypotheses is shown in Table 9.

5 Conclusion

Artificial intelligence (AI) is an emerging field of computer science, and its impact on the financial sector and economy are increasing by the passage of time. The use of AI in the financial sector is at its initial stage. As the financial sector adopts AI technology to take advantage of it, different issues like data security, ethical issues, impact on the economy and bottom line of financial statements of companies will arise.

The implementation of AI and other technologies in the GCC region is gaining attention. However, only in recent years, financial institutions have realized the scope of these technologies. AI is influencing the banking sector of GCC and has vast potential due to the presence of high-value-added capability and efficiency of this technology in the economy.

Moreover, issues related to human knowledge, attitude, and perceptions are important considerations that needing investigation. The humans regard AI is a competitor to them as well as a threat to their job security. This premise can be more prevalent in GCC countries where an aggressive drive is underway towards AI adoption. On the other hand, many part of labour force are from low literate countries like India, Pakistan, and Sri Lanka. Therefore, knowledge, attitude, and perceptions of employees affected by AI cannot be ignored at all.

We studied the humanistic aspect of employees dealing with AI in GCC countries. In the survey, respondents were asked about their AI familiarity and awareness from a general business and finance and data science perspectives. The findings show that the overwhelming majority of the participants are familiar with AI from a business and finance perspective. The research also shows that an overwhelming majority of the respondents in the GCC countries are concerned about ethical, security, and data privacy issues.

In terms of policy implications, the financial institutions can adopt AI by the passage of time and these companies have to train their employees

Table 9 Summary of constructed hypotheses

Hypothesis 1:	*Financial sector professionals' perception level about AI knowledge is not above-average*
Hypothesis expression	$\mu_{\text{AI knowledge}} - 0.5 = 0$
Result	$\mu_{\text{AI knowledge}} - 0.5 \neq 0$
Conclusion Reject null hypothesis	Financial sector professionals perception level about AI knowledge is above-average
Hypothesis 2-a:	The country, education level, and financial sector do not have a significant impact on the financial sector experience level
Hypothesis expression	$\mu_{\text{country}} - \mu_{\text{Education}} = \mu_{\text{country}} - \mu_{\text{financial sector}} = \mu_{\text{countryfinanical sector}} - \mu_{\text{Education}} = 0$
Result	$\mu_{\text{country}} - \mu_{\text{Education}} = \mu_{\text{country}} - \mu_{\text{financial sector}} = \mu_{\text{countryfinanical sector}} - \mu_{\text{Education}} \neq 0$
Conclusion Reject null hypothesis	Except country, education level, and financial sector have an impact on financial sector experience level
Hypothesis 2-b:	The country, education level, and financial sector do not have a significant impact on the AI knowledge level
Hypothesis expression	$\mu_{\text{country}} - \mu_{\text{Education}} = \mu_{\text{country}} - \mu_{\text{financial sector}} = \mu_{\text{countryfinanical sector}} - \mu_{\text{Education}} = 0$
Result	$\mu_{\text{country}} - \mu_{\text{Education}} = \mu_{\text{country}} - \mu_{\text{financial sector}} = \mu_{\text{countryfinanical sector}} - \mu_{\text{Education}} = 0$
Conclusion Do not reject the null hypothesis	The country, education level, and financial sector do not have an impact on the AI knowledge level
Hypothesis 3:	The financial sector experience level does not have a significant impact on the level of AI knowledge
Hypothesis expression	$\beta_{\text{Finanical sector experience level}} = 0$
Result	$\beta_{\text{Finanical sector experience level}} = 0$
Conclusion Do not reject the null hypothesis	The financial sector experience level does not have an impact on the level of AI knowledge
Hypothesis 4-a:	The financial sector experience level does not have a significant effect on the optimism level of the use of AI in the financial industry in the future
Hypothesis expression	$\beta_{\text{Future Expectation}} = \beta_{\text{Positive Impact on economy}} = \beta_{\text{Work efficiency and transparency}} = \beta_{\text{Ethical Issues}} = 0$

(continued)

Table 9 (continued)

Hypothesis 1:	*Financial sector professionals' perception level about AI knowledge is not above-average*
Result	$\beta_{\text{Future Expectation}} = \beta_{\text{Positive Impact on economy}} = \beta_{\text{Work efficiency and transparency}} = \beta_{\text{Ethical Issues}} \neq 0$
Conclusion Reject null hypothesis	The financial sector experience level has a significant effect on the optimism level of use of AI in the financial industry in the future
Hypothesis 4-b:	AI knowledge level does not have a significant effect on the optimism level of use of AI in the financial industry in the future
Hypothesis expression	$\beta_{\text{Future Expectation}} = \beta_{\text{Positive Impact on economy}} = \beta_{\text{Work efficiency and transparency}} = \beta_{\text{Ethical Issues}} = 0$
Result	$\beta_{\text{Future Expectation}} = \beta_{\text{Positive Impact on economy}} = \beta_{\text{Work efficiency and transparency}} = \beta_{\text{Ethical Issues}} \neq 0$
Conclusion Reject null hypothesis	AI knowledge level does not have a significant effect on the optimism level of use of AI in the financial industry in the future

to handle this new technology. This adoption will have a positive impact on the bottom line of financial statements of the financial institutions due to an increase in efficiency and decrease in management and operating costs. On the other side, governments that play a vital role in transformational changes in the economy must reckon issues and challenges arising from the AI. Therefore, it will be essential to consider these issues while making further advancements in AI in the region.

Thus, technical training of the labour force is essential to meet these challenges in the future. Governments must equip the labour force with modern technology so that proper demand from the consumer side could be maintained. AI presents an excellent opportunity for GCC countries to leverage it for sustainable growth and development.

Finally, it is worth noting some limitations of the current work. This research was based on data collected from professionals working in the financial industry in the GCC region through a survey. The sample size was relatively small and can be expanded in future research. Moreover, the research did not use any macro-economic data like production, imports, exports, to shed more light on the and real impact of AI in the GCC region. AI is still considered as a nascent field in the region, but more research is needed and expected to address the stated gaps.

Acknowledgements The author acknowledges the support of Mr. Ammad UlRehman in the data collection process.

References

Acemoglu, D., & Restrepo, P. (2020). The wrong kind of AI? Artificial intelligence and the future of labour demand. *Cambridge Journal of Regions, Economy and Society*, *13*(1), 25–35.

Alizz Islamic Bank. (2019). *Alizz Islamic bank successfully deploys Sunoida's vision—Banking Business Intelligence (BI) and data analytics platform* [Online]. Available from: http://alizzislamic.com/Media-Centre/Press-Releases-Details/snmid/628/snmida/631/snid/190/sname/Alizz-Islamic-Bank-successfully-deploys-Sunoida-s-Vision-Banking-Business-Intelligence-BI-and-Data-Analytics-Platform. Accessed 18 June 2019.

Alzaidi, A. A. (2018). Impact of artificial intelligence on performance of banking industry in Middle East. IJCSNS *International Journal of Computer Science and Network Security, 18*(10), 140–148.

Argam. (2019). Saudi investment bank signs digital deal with SAP [Online]. Available from: https://www.argaam.com/en/article/articledetail/id/599536. Accessed 18 June 2019.

Augustine, B. D. (2018). *Middle East's banking industry headed for tech revolution* [Online]. Available from: https://ifnfintech.com/islamic-bank-becomes-the-first-in-kuwait-to-integrate-ai/. Accessed 30 June 2018.

Bogale, T. E., Wang, X., & Le, L. B., (2018). *Machine intelligence techniques for next-generation context-aware wireless networks* [Online]. Available from: https://arxiv.org/abs/1801.04223. Accessed 18 June 2018.

Davenport, T., Guha, A., Grewal, D., & Bressgott, T. (2020). How artificial intelligence will change the future of marketing. *Journal of Academy of Marketing Science*, *48*, 24–42.

Deloitte. (2018). *Artificial intelligence* [Online]. Available from: https://www2.deloitte.com/content/dam/Deloitte/nl/Documents/deloitte-analytics/deloitte-nl-data-analytics-artificial-intelligence-whitepaper-eng.pdf. Accessed 22 October 2019.

Eisazadeh, S., Shaeri, Z., & Ali, B. (2012). An analysis of bank efficiency in the Middle East and North Africa. *The International Journal of Banking and Finance, 9*(4), 28–47.

El Shazly, M. R., & Lou, A. (2020). Modeling diversification and economic growth in the GCC using artificial neural networks. *Journal of Advances in Economics and Finance*, *5*(1).

Global Ethical Banking. (2018). *Kuwait is the past, Dubai is the present, Doha is the future* [Online]. Available from http://www.globalethicalbanking.com/gcc-companies-financial-institutions-embrace-artificial-intelligence-ai/. Accessed 25 June 2018].

Gremm, J., Barth, J., & Stock, W. G. (2015). Kuwait is the past, Dubai is the present, Doha is the future. *Journal of Islamic finance and Business Research, 1*(1), 1–13.

Grorge, J. (2018). *Rakbank to tie-up with three new fintech firms* [Online]. Available from https://www.tahawultech.com/news/rakbank-to-tie-up-with-three-new-fintech-firms/. Accessed 30 July 2018.

Gunning, D., & Aha, D. (2018). 'DARPA's explainable Artificial Intelligence', (XAI) Program. *AI Magazine, 40*(2), 44–58.

IFN Fintech. (2018). *Islamic bank becomes the first in Kuwait to integrate AI* [Online]. Available from https://ifnfintech.com/islamic-bank-becomes-the-first-in-kuwait-to-integrate-ai/. Accessed 30 June 2018.

Jain, S. (2019). *The potential impact of AI in the Middle East* [Online]. Available from https: https://www.pwc.com/m1/en/publications/documents/economic-potential-ai-middle-east.pdf. Accessed: 31 July 2019.

Manning, J. (2018). *How AI is disrupting the banking industry* [Online]. Available from https://internationalbanker.com/banking/how-ai-is-disrupting-the-banking-industry/. Accessed 30 July 2018.

Mannino, A., Althaus, D., Erhardt, J., Gloor, L., Hutter, A., & Metzinger, T. (2015). *Artificial intelligence: Opportunities and risks*. Policy paper by the Effective Altruism Foundation (2): 1–16.

Mashreq Bank. (2018). *Mashreq bank selects blue prism to drive innovation across all banking functions* [Online]. Available from https://www.mashreqbank.com/uae/en/news/2018/february/mashreq-bank-selects-blue-prism-to-drive-innovation-across-all-banking-functions. Accessed on 27 August 2020.

Mata, J., Miguel, I., Duran, R., Merayo, N., Singh, S. K., Jukan, A., & Chamani, M. (2018). Artificial intelligence (AI) methods in optical networks: A comprehensive survey. *Optical Switching and Networking, 28*, 43–57.

Mckenna, B. (2015). *Al Rayan Bank finds business agility in cloud applications* [Online]. Available from: https://www.computerweekly.com/news/4500252667/Al-Rayan-Bank-finds-business-agility-in-cloud-applications. Accessed 20 June 2018.

Pera, E. (2018). *New technologies set to disrupt UAE banking sector in 2018* [Online]. Available from https://www.menaherald.com/en/money/banking/new-technologies-set-disrupt-uae-banking-sector-2018-kpmg-banking-perspectives-report. Accessed: 21 June 2018.

SAS. (2016). *SAS extends partnership with RAKBANK for analytics solutions to deliver efficient marketing strategies* [Online]. Available from: https://www.sas.com/en_sa/news/press-releases/2016/july/sas-extends-partnership-with-rakbank-for-analytics-solutions.html. Accessed on 31 July 2019.

SAS. (2018). *Bank muscat and SAP recognised for best data analytics initiative in the Middle East* [Online]. Available from: https://www.sas.com/en_ae/news/press-releases/local/2019/bank-muscat-and-sas-recognised-for-best-data-analytics-initiative.html. Accessed on 31 July 2019.

Shalf, J. M., & Leland, R. (2015). *Computing beyond moore's law. Computer* [Online]. Available from https://m-cacm.acm.org. Accessed on 31 July 2019.

Sharma, V., & Tiwari, R. (2015). A review paper on IoT & it's smart applications. *International Journal of Science, Engineering and Technology Research*, *5*(2), 472–476.

Shediac, R., & Samman, H. (2010). *Meeting the employment challenge in the GCC: The need for a holistic strategy*. Booz and Co.

Shirish, K., Jayantilal, S., & Haimari, G. (2016). *Digital banking in the Gulf Keeping pace with consumers in a fast-moving marketplace* [Online]. Available from https://www.mckinsey.com/~/media/McKinsey/Locations/Europe%20and%20Middle%20East/Middle%20East/Overview/Insights/Digital%20banking%20in%20the%20Gulf/Digital%20Banking%20in%20the%20gulf%20161116%20DIGITAL.ashx. Accessed: 31 July 2018.

Sophia, M. (2018). Banks are investing massively into IT Services. *Forbes Middle East staff*. Available from: https://www.forbesmiddleeast.com/en/banks-are-taking-note-as-fintech-spikes-customers-interest/. Accessed: 31 July 2019.

Verma, S. (2017). *UAE banking on AI, and the results are showing* [Online]. Available from https://www.khaleejtimes.com/editorials-columns/uae-banking-on-ai-and-the-results-are-showing. Accessed 30 July 2018.

Zawya. (2018). *Bank Muscat and SAS recognised for best data analytics initiative in the Middle East* [Online]. Available from: https://www.zawya.com/uae/en/press-releases/story/Bank_Muscat_and_SAS_recognised_for_Best_Data_Analytics_Initiative_in_the_Middle_East-ZAWYA20180809102646/. Accessed: 18 June 2018.

CHAPTER 7

The GCC and Global Health Diplomacy: The New Drive Towards Artificial Intelligence

Mohammed Sharfi

1 Introduction

Health diplomacy is a new growing area of study endeavouring to approach health in an encompassing global context. There has been a marked intensity in public or academic writings in this field. These writings have taken the multi-disciplinary approach or cross-cutting sectors of the subject. Generally, they focused on two aspects: first, the use by states of health for the implementation of their economic, strategic, and ideological objectives; and second, the global debate on the difficulties of collective action and resolving health problems through cooperation in different sectors and levels (Kickbusch and Behrendt 2017). In recent years, the GCC (Gulf Cooperation Council) countries allocate huge financial resources for their health sectors as part of their overall strategic visions. Health diplomacy plays a critical role in this policy direction with the need for know-how in an intrinsically interdependent world.

M. Sharfi (✉)
Independent Researcher and Senior Consultant, Ministry of Foreign Affairs, Doha, Qatar

E. Azar and A. N. Haddad (eds.), *Artificial Intelligence in the Gulf*,
https://doi.org/10.1007/978-981-16-0771-4_7

The GCC is also planning to be one of the top destinations for healthcare worldwide. The momentum in the health sector is reflected in the different GCC visions such as United Arab Emirates (UAE) vision 2021, Qatar Vision 2030, Saudi Arabia Vision 2030, Kuwait vision 2035, and Bahrain vision 2030 (KSA Government 2019; Qatar Government 2019; UAE Government 2019).

The United Nations General Assembly resolution recognized the close relationship and the interdependence of health and foreign policy; therefore, it became an integral part of the sphere of diplomacy (United Nations 2009). "Global health diplomacy brings together the disciplines of public health, international affairs, management, law and economics" (Kickbuch and Rosskam 2012). This emergent concept strives to illustrate the practices by which governments and non-state stakeholders seek to coordinate policy response to advance global health. There are growing trans-border health issues that prompt the significance of diplomacy in a fast-moving world in the era of globalization. Recent examples of health challenges include the outbreak of Severe Acute Respiratory Syndrome virus (SARS), Ebola pandemics, and the threat of HIV/AIDS. The Oslo Ministerial Declaration of 2006, which brought together foreign ministers of seven countries, argued that global health should hold a strategic place on the international policy agenda and explored how better-linking health concerns to foreign policy could be beneficial in achieving both health and foreign policy goals (Amorim et al. 2007). Some writers introduced a structured analytical approach to this new concept such as using five metaphors as drivers for global health action: global health as foreign policy, security, charity, investment, and as public health (Stuckler and McKee 2008). Others recognized four main discourses on global health governance: "biomedicine, economism, human rights and security" (Lee 2009).

In this context, healthcare is high on the GCC foreign policies agendas as they become more driven in participating and conducting medical research. The financial resources provided by the GCC to introduce technology in the health sector for the transformation of their economies will lead to a positive contribution to global health diplomacy. In a period of a highly volatile political and economic context in the GCC, it is improbable leaders will focus on integrating Artificial Intelligence (AI) in the health sector without considerable engagement. It is certain that AI will change the provision of healthcare in future. Health systems around the world pursue the best possible patient experience. Science and technology play a critical part in advancing these objectives as they transformed other key sectors in modern societies.

Technological breakthroughs will abet healthcare to make services more available, cost-effective, and efficient. "They will help health practitioners diagnose people earlier, treat them faster and more effectively, and provide new opportunities for patient engagement, empowerment and self-care" (WISH Data Science and AI Forum 2018). Scientific studies show constantly the benefits of AI in healthcare such as the study by University of Nottingham that indicate AI can predict the risk of early death due to chronic disease, or the software created by Imperial College London and the University of Melbourne that forecast the survival rates and response to treatments of patients with ovarian cancer (Jefferson 2019). Scientists are also developing AI systems that can diagnose common childhood conditions, predict whether a person will develop Alzheimer's disease and monitor people with conditions like multiple sclerosis and Parkinson's disease (Brookes 2019). This technology can also speed up many aspects of drug discovery and, in some cases, perform tasks typically handled by scientists (Metz 2019). Therefore, observers envisage that the current emphasis on big data in healthcare will persist, with the global market expecting to be worth US$34 billion by 2022 (WISH Data Science and AI Forum 2018).

The GCC commitment to devote significant financial resources in technology, particularly AI, is due to its future dividends in their strategies, along with the abundant financial resources and concrete technology infrastructure. At present, there is significant momentum driving the GCC global health partaking especially in the AI sphere. The trend of the GCC to focus on AI as a strategy is clearly emanates from the domestic agenda, and the need to bridge the technological gap between them and developed countries in this domain. Technological gap is defined "as the differences in technological or knowledge advancements between firms or countries" (UNCTAD 2014).

Although AI is in early stages, it enjoys a lot of interest and the world, including the GCC, is already garnering the advantages of progress made in AI technology. Therefore, the adoption of this technology in the healthcare field became an important dimension in the health diplomacy of the Gulf region. The need for GCC states to collaborate with leading international actors in AI is an integral part of their pursuit for the post-hydrocarbon era, and break away with its long dependency in the energy sector. Many countries in the region recognized the prospects of AI.

Therefore, they became early adopters of this technology and established the region as a global leader in this domain (Elsaadani et al. 2018). In this framework, the GCC have enacted national policies and supported international and multilateral agreements and deliberations aimed at boosting and promoting AI integration in their health sectors. The prospect of AI application in healthcare in the GCC depends not only in government to government collaboration but also on the private sector. Therefore, the GCC international cooperation on AI and Health is taking a variety of form. Diplomacy plays a critical part in the GCC foreign policy by facilitating this policy agenda.

This chapter will look at the health diplomacy concept, then provide an insight into its emergence in the GCC, and finally discuss the GCC Drive towards AI partnerships.

2 Health Diplomacy: The Concept

There are different approaches to define health diplomacy in the literature; however, the overarching approach to this topic discusses methods and ways to further deepen the link between global health and foreign policy agendas. These definitions focus "on the field being driven by globalization, diverse actors beyond nation-states, health negotiations, health impact of non-health negotiations, and most importantly the normative goal of using foreign policy to support global health" (Feldbaum et al. 2010). As for the Global Health Diplomacy (GHD) term, there is no particular definition that encompasses its scope. One of the definitions of GHD refers to "both a system of organization and to communication and negotiation processes that shape the global policy environment in the sphere of health and its determinants" (Kickbusch and Kokney 2013). Another definition referred to it as "international diplomatic activities that (directly or indirectly) address issues of global health importance, and is concerned with how and why global health issues play out in a foreign policy context" (Michaud and Kates 2013).

This is a vital dimension for the state since health impacts its economy and national interest; therefore, there is a need for new capacities in internal systems. In 2009, the UN Secretary-General Ban Ki-moon stated "global health touches upon all the core functions of foreign policy: achieving security, creating economic wealth, supporting development in low-income countries and protecting human dignity. Government and non-government stakeholders have started to recognize the strategic value of how and why the foreign policy community's support for the health sector is vital for advancing both. The need for increased foreign policy and diplomatic activities on global health problems has created opportunities and challenges for those who shape the foreign and health policies of member states" (UN General Assembly 2009). This statement reflects the change in the nature of health issues from its local confines to the global context.

There are broad issues related to health that go beyond the state borders necessitate global collaboration to address them. Therefore, the link with diplomacy is centred on its capacity to negotiate and attain applicable solutions, and this is critical in the globalization era. Another definition is that "health diplomacy is the chosen method of interaction between stakeholders engaged in public health and politics for the purpose of representation, cooperation, resolving disputes, improving health systems, and securing the right to health for vulnerable populations" (Health Diplomats 2019).

The wide-ranging impact of health issues in the social, economic, and political arenas prompted diplomacy to involve in addressing these complications, while public health practitioners engage in diplomacy. Diplomacy became an integral part of handling partnerships in the field of global health that link stakeholders in different fields. "Simple classifications of policy and politics — domestic and foreign, hard and soft, or high and low — no longer apply" (Kickbusch 2003). In this context, Kickbusch and Kokney (2013) indicated that GHD "if well conducted, results in improved global health, greater equity, better relations and trust between states and a strengthened commitment on the part of stakeholders to work together to improve health nationally and globally". The World Health Organization (WHO) indicates that it "brings together the disciplines of public health, international affairs, management, law, and economics, and focuses on negotiations that shape and manage the global policy environment for health" (WHO 2018).

The increasing linkages between countries at the international level led to substantial bearing on the national health agendas and will continue to influence domestic policies in this sector. Health challenges transcend international boundaries and could lead to destabilization of global order. This context provides a platform for cooperation between governments and the private sector to address collective action problems such as disease epidemics and climate change. "The recognition of the need for policy coherence, strategic direction and a common value base in global health is only just beginning to emerge at the level of nation-states" (Kickbuscha et al. 2017).

The rise of soft power as a concept in foreign policy is another dimension that prompts states to adopt health diplomacy as a tool in their interaction with the outside world. Soft power is exercised through the promotion of values, culture policies, and institutions while hard power is concerned with military and economic resources (Nye 2005). "Medical diplomacy" became part of the foreign policy of many states such as Cuba which adopted this approach since its revolution. Cuba is a major contributor in filling the gap of the brain drain to the west in remote and neglected parts of the world and building a positive image through this role. In the past half-century, some 130,000 Cuban medical personnel have worked abroad. Today, the number of personnel is 37,000, spread across more than 70 countries; half of them are doctors, while the rest are nurses and other specialists (Jack 2010).

Globalization propelled the link between foreign policy, health, and international trade to the forefront of multilateral diplomacy. There is a growing understanding that the global health field has become multifaceted with an increased number of stakeholders' whether state or non-state actors. The domestic context and global health issues are interdependent and their dynamics call for policy coherence and improved coordination mechanisms between the state diplomatic apparatus and global health forums.

3 Health Diplomacy in the GCC: An Insight

According to the WHO, the world spent US$7.5 trillion on the health sector in 2016, which represents almost 10% of global GDP (WHO 2018). Healthcare is one of the fastest expanding sectors in the global economy, and spending will increase in countries all over the world. According to Deloitte, the growth will be driven by the needs of ageing and growing populations, the prevalence of chronic diseases, emerging-market expansion, infrastructure improvements, and advances in treatment and technology (Deloitte 2015). The Middle East and Africa will witness the largest spending increase in healthcare. The GCC spending in the health sector is expected to reach US$69.4 billion in 2018, with continuous growth in population, growing incidence of chronic conditions, numbers of, and increasing healthcare costs (Deloitte

2015). The emergence of strong and persistent government commitment in the GCC for progressive healthcare linked to technology derived from the committed leadership for this policy agenda. The political will and individual leadership in the GCC were instrumental in galvanizing commitment in engaging with health diplomacy. However, there were divergences in various Gulf countries in terms of their strategies and policies towards their engagement focus in health diplomacy.

Global health diplomacy is critical for the GCC in its endeavour to strengthen health services and overhaul their economies. It is part of the public policies in the region geared towards modernizing the delivery of services with the available cutting-edge technologies which will lead to structural changing in their economies. GHD advance the national interest of the GCC states with the transfer of technology and knowledge sharing constitute an important component in building competitive health systems. The procurement of technology in the international sphere is a long-term and multifaceted endeavour that demands policy continuity, organization, and integration. Therefore, the greater focus of the GCC on changing their economies translates into concerted interest in the question of transfer of technology and the unevenness of the international distribution of technological and innovation capabilities. The Gulf region can benefit from the AI technology opportunities in the international sphere through platforms such as conventions that reflect share needs, innovations, and good practices. This trend can lead to prospects of collaborations that can improve health outcomes since embedding technology in the health sector became a solid area of policy concern in the Gulf.

The healthcare spending in the Middle East and North Africa (MENA) is forecast to rise to US$144 billion by 2020, with GCC nations alone contributing around 52% of the total healthcare expenditure (Al Masah Capital 2014). A total 707 GCC healthcare projects worth more than US$60.9 billion are currently under development “264 projects worth US$24.7 billion are under construction, 227 projects worth US$12.7 billion are in the design and pipeline, while there are 75 healthcare projects with a combined estimated value of US$1.76 billion (AED6.46 billion) are in tender phase” (Pereira 2018). The huge funding allocated to the health in the GCC reflects the direction towards global collaboration in the development of this sector in the region. Global health diplomacy plays a critical role in this context with the GCC delves into collaboration in a wide range of issues related to healthcare. Many GCC countries are determined to accomplish global health standing through hosting of high-profile international conferences and discussion forums. The GCC countries are also engaging in research to find better solutions to their healthcare problems. In general, they view partnership in the multilateral, bilateral levels as well as with cutting-edge tech companies as an increasingly significant foreign policy and soft power instrument.

In Qatar, the interest on scientific research in the field of health is reflected on the presence of about 20 independent institutions that conduct research in the fields of health, such as Sidra Medical Center, Qatar Bionic Medical Research, Qatar University, Qatar Institute for Biomedical Research. These research centres in Qatar include a group of scientists from all around the world who are focused on exploring solutions to serious medical problems. Qatar also launched the World Innovation Summit for Health (WISH), a global initiative under the umbrella of the Qatar Foundation for Education, Science, and Community Development, to stimulate and disseminate innovation and adopt best practices in healthcare through global cooperation. WISH is led by global experts in the areas of healthcare innovation, focusing in areas such as diabetes, child mental health, dementia, healthcare, data science, and AI.

Another ambitious initiative in the field of medicine is Sidra Medical and Research Center with a 7.9 billion fund from Qatar Foundation. The Sidra Center will focus on paediatrics and women's health issues and raise the status of the State of Qatar to be a leader in the field of scientific research. The centre will employ more than 5,000 people, including 600 doctors and 2,000 nurses, attracting world-class staff in the field of research and education (Roberts 2014). In joint projects between Qatar and other countries, a Cuban hospital was opened in January 2012 with 400 Cuban doctors and nurses, and a Turkish hospital in February 2017.

In Saudi Arabia, the government collaborated with WHO in the area of disease control in relation to the annual religious pilgrimage to Mecca. This annual gathering brings people from 180 countries, with an estimated 2 million people. The WHO Collaboration Centre for Mass Gatherings Medicine, as well as, US centres for disease control provided technical expertise for Saudi Arabia in area. There were 25,000 health workers deployed to the eastern province for the pilgrimage to treat the sick and collect epidemiological data "in addition to permanent hospitals in Mecca and Medina, about 25 temporary hospitals and clinics with over 5,000 hospital beds are opened every year" (Craig 2018). This global research collaboration in Hajj provided an invaluable insight into infectious diseases.

The UAE launched Reaching the last mile, which entails cooperation with international partners of activist-philanthropists, Non-Governmental Organizations (NGOs), pharmaceutical, and academic partners to combat neglected tropical diseases; these include polio, Guinea worm disease, lymphatic filariasis, river blindness, and malaria. The UAE government set up the Reaching the Last Mile Fund, which is a 10-year, US$100 million fund launched in 2017 by His Highness Mohamed bin Zayed, Crown Prince of Abu Dhabi, and supported by the Bill & Melinda Gates Foundation, Department for International Development (DFID), United States Agency for International Development (USAID) and the ELMA Philanthropies (Reachingthelastmile.com 2019). This initiative will have a great impact in addressing such diseases in underdeveloped countries, and provide an important example of the application of health diplomacy through different stakeholders.

In 2015, the renowned US Cleveland Clinic was opened Abu Dhabi, as part of Mubadala's network of world-class healthcare facilities with medical experts from all over the world. The UAE Mubadala established a Telemedicine Centre in a joint venture with Switzerland's leading telemedicine provider, Medgate (INSEAD 2019). The government also collaborated with institutions such as South Korea's Wooridul Hospital to set up Healthpoint, a world-class speciality hospital that houses the Wooridul Spine Center; and partnered with Imperial College London to establish the Imperial College London Diabetes Centre (Mubadala 2019). There are various collaborations between the Ministry of Health and U.S. Medical Equipment Companies (such as GE, Healthcare, 3 M, Abbott, Medtronic, and Johnson & Johnson) and also Electronic Health Record and Health Information Exchange (such as Cerner Corporation, Epic Systems, InterSystems) (U.S.-U.A.E. Business Council 2018). The UAE's established healthcare system, combined with its international standing as a tourist destination, has propelled the growth of its medical tourism. Dubai became an important medical tourism hub with a reasonable cost of treatment with cutting-edge medical facilities along with the leisure aspect. The UAE is one of the fast-growing medical tourism hubs as the medical tourism in Dubai generated more than AED 1.4 billion for the emirate in 2016 and that the city received 326,649 medical tourists according to Dubai Health Authority (DHA) (Government of Dubai 2016). According to Patients Beyond Borders, the global medical tourism market is estimated to be around US$65 billion to US$87.5 billion based on approximately 20-24 million cross-border patients worldwide (Patients Beyond Borders 2019). The GCC, especially the UAE, could take a good share out of this huge market through the use of technology. The UAE healthcare market is projected to grow by 60% in the next five years to AED103 billion (US$28 billion) by 2021 (MENA Research Partners 2017).

4 The GCC Drive Towards AI Partnerships

The future healthcare will require the use of AI in facilitating the delivery of an efficient system since it became an integral part of all spheres of life. AI impacts the provision of healthcare delivery positively. "It helps to improve the operational performance and efficiency of workflows, supports high-quality and integrated clinical decision-making, empowers patients and consumers to manage their own health proactively, and enables population health management" (Fidanci 2019). This will be through different means such as the use of this technology in diagnostics, medication of illnesses, remote operations in certain settings and decision-making. "Unlike humans, AI-based machines are unbiased. For example, once it is appropriately trained, an AI system may give a senior leader a very unbiased, unblinkered view of the information and possible options, and the individual can then decide on what actions to take" (The Economist Intelligence Unit 2016). The model, which entails connecting AI with a large collection of data, could provide vital insights into trends and patterns which will transform the delivery of healthcare.

According to a survey conducted by PWC, the Middle East embrace the adoption of electronic health records faster than many parts of the world, and this provides huge opportunity to collect data and using AI for analysis and application (PWC 2017). The report stipulates the trend could potentially see this region leading the way in this important healthcare trend. "close to half and up to 73% of all respondents are willing for a robot to perform a minor surgical procedure instead of a doctor. This willingness ranged from 62% for the UAE to 55% in Saudi and 65% in Qatar" (PWC 2017). This will open the door for wide-ranging partnership between the GCC and global public and private sectors in the AI field. The Riyadh Global Digital Health Summit held in August 2020 indicated AI has been critical in controlling the spread of the coronavirus in the GCC. Technology has forecast the pandemic's development and informed residents when they have been in contact with infected individuals (Bradsley 2020). AI's capacity to analyse huge quantity of data enabled authorities all around the world to gather information in their efforts to halt the pandemic. The Saudi Authority for Data and Artificial Intelligence launched an application that uses the "Exposure Notification" technologies recently developed by Apple and Google to help alert users who may develop symptoms as a result of contact with people who have been diagnosed with COVID-19 disease (Salama 2020).

The huge budgetary provision for the healthcare sector in the GCC entails the use of new technologies and the implementation of innovative related systems. GCC states are heading towards implementing the latest healthcare technology practices in their pursuit to transform their hydrocarbon economies. They also seek to strengthen their competitiveness and standing in the international level, while increasing their significance in the adoption of technology. According to Accenture analysis, AI has the potential to significantly raise economic growth rates in the region, adding US$215 billion and US$182 billion in annual Gross Value Add (GVA) to the economies of Saudi Arabia and UAE, respectively, by 2035 (Elsaadani et al. 2018). To achieve these objectives, the GCC needs to collaborate with leading international actors in the field of technology and AI. Private healthcare providers will also play a central role in their partnership with the GCC to deliver cutting-edge healthcare. Saudi Arabia did not yet publish official AI strategy, while the UAE, accomplished some of the objectives in its 2016 National Transformation Strategy. It includes the application of AI-driven tools, in its goal of using IT to improve the efficiency and effectiveness of healthcare. Saudi Arabia is also boosting investment in its digital economy as part of the objectives of the national Vision 2030 development programme which also fit with the drive towards AI (Economist Intelligence Unit 2018). In 2019, the UAE launched the first-graduate level, research-based Artificial Intelligence (AI) university in the world, Mohamed bin Zayed University of Artificial Intelligence (MBZUAI) based in Abu Dhabi. The university, which opens in August with 75 masters and PhD students, will focus on key applications of AI, such as computer vision, which helps to detect cancers, machine learning and natural language processing (Rose 2020).

In Qatar, the significance of adopting AI is also related to the implementation of its 2030 vision of transforming the state into a knowledge-based economy. According to the Medical Tourism Index, Qatar is ranked 30th in the world for medical tourism with significant investments made in its healthcare infrastructure; this includes new hospitals with prominent international accreditation (Medical Tourism Index 2019). The country's healthcare spending is among the highest in the Middle East, with QR22.7 billion invested in healthcare in 2018, a 4% increase from the previous year (Gulf Times 2019). In this context, the National Artificial Intelligence Strategy for Qatar, a document prepared by Qatar Computing Research Institute (QCRI), intends to inform and advise Qatar's leadership on AI pointed it would benefit accelerate expansion in medical tourism (Qatar Computing Research Institute 2019). However, the report indicated that one of the elements impact the cost of adoption of AI in Qatar is its capacity to import cheap labour from abroad. "The cost of labour-intensive and inefficient solutions is often less than the cost of an AI solution leading to a situation where long-term efficiencies are sacrificed for short-term benefits" (Qatar Computing Research Institute 2019). Due to various aspects, Qatar is well-positioned to take advantage of this golden opportunity and become a critical player in an AI-focused future economy.

The Qatar Genome Project provides fine-grained data about citizens of Qatar and should serve as the basis of uncovering persistent disease patterns in the local population. The project is generating large databases combining whole genome sequencing and other omics data with the comprehensive phenotypic data collected at Qatar Biobank. The wealth of such data empowers researchers to make breakthrough discoveries as well as help policymakers to better plan for future healthcare directions in Qatar. The project is collaborating with local and international stakeholders with solid bioinformatics research bases, such as Hamad Hospital, SIDRA, Weil-Cornell, Qatar Biomedical Research Institute, and QCRI. In this context, Qatar Genome Project signed a strategic research and development agreement with Genomics England, aimed at enabling novel scientific discoveries and providing medical insights in genomics and precision medicine (Qatar Genome 2019).

In this context, in line with Saudi Vision 2030, Saudi Arabia's Ministry of Health signed a Memorandum of Understanding (MoU) last year with General Electric (GE). The partnership focuses on enabling and accelerating Digital Transformation of KSA's health sector, in which GE would provide advanced technical solutions such as health system control centres, simulation centres, virtual hospitals, improved patient experience in health facilities and extraction of cost-efficiencies (Arab News 2017).

According to WHO, Saudi Arabia's public health sector is 'overwhelmingly financed, operated, and monitored by the Ministry of Health (MoH)'. With 2,390 healthcare centres and 279 hospitals there are significant needs to be met (Nagel 2019). Foreign pharmaceutical manufacturers are actively encouraged to establish plants in Saudi Arabia through public-private partnerships and joint ventures with national entities. Under the National Transformation Program (NTP) investment plans, the Saudi Ministry of Health has been allocated a budget of SAR23 billion (US$6 billion) in the use and advancement of IT and digital transformation (Naseba 2019). The objective is to improve the efficiency and effectiveness of the healthcare sector through the use of technology and digital transformation.

Saudi Arabia embarked on extensive investment on AI as part of its future, as the Crown Prince announced in October 2017 a new planned US$500 billion, 26,500 square Km mega-city, in the northwest desert is also part of the government Vision 2030. The crown prince, Mohamad bin Salman al-Saud, is interested in technology and AI stating in October 2017 during the Future Investment Initiative regarding the new smart city "Everything will have a link to artificial intelligence, to the internet of things". In the same event, Saudi Arabia granted the citizenship to Sophia a human-robot developed by Hanson Robotics. Saudi Arabia is also preparing for the post-oil with its extensive investments in technology. The Saudi government committed a US$45 billion to SoftBank Group Corp's US$100 billion technology fund and other billions of dollars invested in companies such as Tesla, Uber, and other tech startups (Sharif 2018). PwC estimates that the shift towards AI Technology will contribute in few years to 12.4 per cent of GDP in Saudi Arabia, and 13.6 per cent of the GDP in the UAE and 8.4 per cent of the GDP in the other four GCC countries (PWC 2018).

The National Digitization Unit (NDU) and the Future Investment Initiative are part of Saudi Arabia drive towards transforming the state to the digital era and AI (NDU 2019). The NDU aims to boost the role of the digital world in the country and encompasses various initiatives and public-private partnership, while the Future initiative mandated with identifying industries to be transformed by AI. Many multinationals such as the US General Electric and SAP partner with NDU to achieve this objective (Al Bilad 2018; Saudi Gazette 2017). The Saudi Public Investment Fund committed US$45 billion to the Japanese technology investor, Softbank Vision Fund in AI-related technologies (Economist

Intelligence Unit 2018). The Vision Fund's largest investments so far are in two microprocessor firms, ARM and NVIDIA, which are developing some of the hardware being utilized for AI and robotics. It has also invested in firms such as Brain Corp, which is a development software for autonomous robots (Economist Intelligence Unit 2018). Meanwhile, Saudi Arabia is partnering with leading technology multinationals such as GE, SAP, and Google to develop its capacities. The Saudi National Initiatives in AI and technology health-related issue necessitate extensive collaboration with global leaders in this field.

The UAE, in particular, became a regional pioneer in embracing AI technology. Dubai is also developing a plan to be a central hub for in the use of AI in the medical sector, while the UAE appointed in October 2017 a minister for AI the first time in the world. In Saudi Arabia, a similar appointment was made in November 2017, with Ahmed Altheneyan becoming the first deputy minister for technology, industry and digital capacities at the Ministry of Communications and Information Technology (MCIT) (Economist Intelligence Unit 2018). Undoubtedly, these steps will propel the significance of AI use in healthcare in line with the GCC states visions. Saudi Arabia and UAE could play a central part in the Gulf region drive towards the embrace of AI as a result of their large Information Technology market and a strong economy. The GCC will continue to utilize AI to improve healthcare sector efficiency, significantly cut reliance on expatriate workers and improve the quality of its citizens‘ lives. The current GCC countries in particular UAE and Saudi Arabia global health strategy underline far-reaching and commitment to utilize technology particularly AI in their health sectors. There are multiple potential collaborations in the area of AI in healthcare, in which the international community can develop through governmental or private partnerships.

Health is one of the key sectors in the UAE's Artificial Intelligence strategy released in October 2017 intended to achieve the its Centennial 2071 objectives. The focus in AI is a critical dimension in shaping the future of the UAE and embraced by the decision-makers at the highest level. His Highness Sheikh Mohammed Bin Rashid Al Maktoum Vice President and Prime Minister of the United Arab Emirates, and Ruler of Dubai stated "we want the UAE to become the world's most prepared country for artificial intelligence" Dubai aims to be the leading hub for cutting edge in the field of health Artificial Intelligence (Salama 2017). The UAE formed the Council for Artificial Intelligence to ensure the implementation of these technologies in the various state sectors. The UAE became a hub for international events that debate AI and healthcare issues such as the annual Dubai health forum, which features specialists and experts in the field, and the World Government Summit (WGS). "The World Government Summit is a knowledge exchange center at the intersection of government, futurism, technology, and innovation. It functions as a thought leadership platform and networking hub for policy-makers, experts and pioneers in human development" (KPMG and World Government Summit 2019). Dubai continues to play a significant role in global health in comparison with its small size.

Dubai organized Artificial Intelligence Week Middle East in September 2018, which included an emphasis on the health sector. This was the first time AI event in the Middle East, and its main theme was "Personalised and predictive: the new era for AI-powered patient care". It included over 400 attendees with talks from the regional and international pioneers on the different applications for AI in healthcare. This event, which convenes on an annual basis, reflects the dedication of the UAE government towards AI integration into the health sector (AI Intelligent Automation Network 2019).

In this context, the UAE Ministry of Health and Prevention has become a worldwide pioneer in the use of a smart application to aid treat chronic or recurrent depression, or multiple depressive disorder, associated with Multiple Sclerosis, MS, for patients. The DePrexis MS application utilizes AI to simulate functional interventions based on psychotherapy research. The app uses the Deprexis technology developed by German company Gaia AG, which provides support to sufferers of depression through smartphones, tablets, or PCs (Emirates247 2018).

DHA also circulated the results of the Chest X-ray AI Algorithm used through DHA medical fitness centres. This joint collaborative effort with Agfa HealthCare and in partnership with VRV began developing Machine Learning enabled detection of abnormal Chest X-ray finding. The collaboration is the first Augmented Intelligence (AI) validation in the UAE. As part of the Algorithm processed approximately 4,900 Chest X-rays, and two DHA MFC Radiologists reviewed the findings detected by the AI Algorithm. The two Radiologists provided feedback via an AI-enabled workflow if they agreed or disagreed with the AI algorithm findings (Dubai.ae 2018).

Another indication of the UAE continuous focus on technology is Dubai Future Accelerators (DFA), a programme introduced by the

Dubai Future Foundation, and the Government of Dubai that invites entrepreneurs to submit proposals for emerging technologies (Dubai Future Accelerators 2019). This large programme gathers in Dubai leading international companies and entrepreneurs to tackle key "21st century opportunities". This includes the application of the forefront technologies like genomics, robotics, and AI.

DHA has signed MoUs with four cutting-edge international companies to adopt their inventive healthcare innovations in AI (Government of Dubai 2018). The health authority is implementing virtual health through an application in partnership with Babylon, which uses AI technology to provide remote General Practitioner, GP, consultations round-the-clock (Arab Health Online 2018). It also partners with Healthcare and Innovative New Technology (HiNT) in their innovative stroke detection headband. HiNT "developed a wearable point-of-care monitoring device that detects when patients at high-risk are having a stroke. The device alerts the caregiver, the ambulance and the emergency within minutes" (HiNT Neuro 2019). DHA also planning in association with Bodyo to set up free to use AI pods across Dubai that will do quick health scans for the public and give immediate results (Chaudhary 2018). The fourth company to collaborate with DHA is Admetsys which is developing flow cell sensors that detect sudden drops in vitals in ICU patients through an algorithm that measures these vitals constantly and it can be read by a nurse on the monitor at a glance (UAEBARQ 2018). DHA is already using AI to assist its procurement and contract management process (Government of Dubai 2019).

The UAE Ministry of Health also partnered with Philips Electronics Middle East & Turkey to establish the remote critical care project Tele-ICU, which offers remote, round-the-clock medical consultancy services in the areas of critical care, medical simulation and telemedicine (Ministry of Health and Prevention 2016). The UAE has also signed a Memorandum of Understanding with India to explore new ways of collaboration to facilitate the growth of their AI initiatives. This partnership is estimated to generate economic benefits for both the UAE and India over the next 10 years of up to $20bn (Gulf News 2018). Local investors in Dubai are collaborating Health-tech start-ups such as Singapore's AEvice health, working to develop a wearable device to diagnose asthma earlier, as well as Argentinian OTTAA, which is using predictive algorithms to help speech-impaired people communicate through images (Deloitte 2019). Most of the radiology departments in UAE hospitals will rely on Artificial Intelligence in 2025, which will open wider fields of diagnosis and reduce the error rate (The Arab Hospital Magazine 2019). All these efforts stemmed from the sense of urgency to restructure and diversify their economies as part of their massive modernization plans. Their long-term future depends on reaching out to investors and taking advantage of the new emerging innovations in AI technology. The UAE is one of the most equipped countries in the region to adopt AI technology due to the exhaustive technology infrastructure established by its government over the past two decades.

The UAE venture for future AI integration in its institutions, the Minister for Artificial Intelligence and Invest India signed a Memorandum of Understanding for the India-UAE Artificial Intelligence Bridge. The corporation will create an estimated US$20 billion (Dh73.4 billion) in economic benefits in the next 10 years and will pursue to gauge the dynamic nature of innovation and technology (Gulf News 2018). This partnership will be implemented through arranging a UAE-India AI Working Committee between the UAE Ministry for Artificial Intelligence, Invest India, and Startup India. This joint committee will convene once a year and delegated to raise investment in AI start-ups and research activities in partnership with the private sector.

Artificial Intelligence will have considerable economic impact by boosting productivity worldwide. However, many jobs are at risk due to the magnitude of AI and automation expansion across a range of sectors. This will prompt far-reaching changes in the Gulf region labour market and employment structure since it will affect foreign workers and citizens in sectors with a high risk of disruption by new technologies. According to a McKinsey, nearly 45 per cent of these jobs are technically automatable today, and AI could lay off 2.8 million full-time employees in the GCC, and help save around $366.6 billion in wages (McKinsey 2018) The study indicated jobs in sectors, such as manufacturing, transport, and warehousing, where routine tasks are the norm, are most at risk of being replaced by AI. While those involving tasks with human interaction like the arts, entertainment, recreation, healthcare, and education can breathe easier with a lower automation potential ranging from 29 to 37 per cent.

In the meantime, the consultancy firm PWC has estimated that by 2030, AI could create 10 million new jobs only in the Gulf region (PWC 2018). This outlook is a reflection of a report released by the World Economic Forum, concluded that while automation and AI could displace 75 million jobs globally by 2022 it could also create 133 million new ones

(BBC 2018). Therefore, the GCC countries will have to develop policies, standards, and regulatory AI frameworks to create a balance between the advantages and potential disruptions of AI in the society.

5 Conclusion

Health diplomacy is part and parcel of the "new diplomacy" refer to changes in foreign policy direction and the new dynamics in diplomatic practice. Global change is creating "new collective action problems (e.g., climate change, population migration, economic instability, disease pandemics), for foreign policy and, consequently, diplomats. The new diplomacy agenda thus includes a broader range of issue-areas deemed relevant to foreign policy" (McInnes and Lee 2006). This has led to the emergence of new spheres of diplomacy such as internet diplomacy (negotiation of rules governing the internet) and disaster diplomacy (understanding and addressing risks in a complex global system) and health diplomacy (Comfort 2000).

The GCC, whose economies until recently function far from the technological frontier, endeavour to have a greater focus on public policies that promote a renewed interest in the question of transfer of technology. The GCC policy direction towards embracing AI with the political commitment behind it, whether in public service, academia or, the private sector will enhance the policy effort to strengthen institutional delivery. This will create an environment that attracts investment and international partnership in fostering technological advancements and AI in various GCC public service sectors. The GCC diplomatic engagement on health issues propels global health diplomacy to assume a more prominent role in addressing health strategies in the framework of their ongoing economic visions.

The GCC continues to channel a substantial share of its financial resource into the global health enterprise. They have to work with partners around the world to achieve their national visions and started to contribute considerably to the introduction of AI in their health sector through global cooperation. The GCC is taking fittingly cautious and measured strides towards the emergence of AI in healthcare. The GCC succeeded in managing and collaborating with external stakeholders and enabled the consolidation of the use of AI in the health sector. This engagement reflects the GCC openness for the international community to offer expertise and ideas in the GCC health sector. Global health cooperation particularly in the realm of technology derived from health diplomacy which is an important dimension in the GCC national interest. Major partnerships are taking part between the region and stakeholders in the international health sector particularly pioneers in health technology. The need for the GCC to engage in global partnership and collaboration remain critical in the health sector particularly in AI. Unquestionably, all stakeholders need each other to develop and nurture the application of AI in the health sector. The Gulf region's quest to develop solid healthcare systems and innovations provides ample opportunity for global partnerships in the field. The political leadership in the GCC, particularly the UAE and Saudi Arabia, has a strong conviction that their future lies in the use of AI technology in all sectors.

References

AI Intelligent Automation Network. Artificial intelligence week Middle East. Available at: https://bit.ly/2KSjdU8.

Al Bilad, KSA progressing on the road to Digital Transformation with SAP, 03 November 2018. Available at: https://bit.ly/2HARkfq.

Al Masah Capital, MENA Healthcare Sector Report, 2014, Dubai. Available at: https://bit.ly/2JBBl3q.

Amorim, C., Douste-Blazy, P., Wirayuda, H., Store, J.G., Gadio, C.T., Dlamini-Zuma, N., Pibulsonggram, N., 2007. Oslo Ministerial declaration—Global health: A pressing foreign policy issue of our time. Lancet 369 (9570), 1373e1378. Available at: https://bit.ly/2WtAq9H.

Arab Health online, Dubai to Deploy Cutting-Edge AI devices in Healthcare, issue 5, 2018. Available at: https://bit.ly/2JCpouP.

Arab News, Ministry of Health and GE deliver digital solutions to transform Kingdom's health care sector, 17 March 2017. Available at: https://bit.ly/2YDatE7.

BBC, WEF: Robots 'will create more jobs than they displace', 17 September 2018. Available at: https://bbc.in/31FHi82.

Bradsley, D. AI helped limit spread of Covid-19 in the Gulf, experts hear, The National, 13 August 2020. Available at: https://bit.ly/3jENTpx.

Brookes, D. How artificial intelligence can save your life, The New York Times, 24 June 2019.

Chaudhary, S.B. DHA launches all-in-one health app at Gitex, Gulf News, 14 October 2018. Available at: https://bit.ly/2wdbMh3.

Comfort LK. 2000. Disaster: Agent of diplomacy or change in international affairs. Cambridge Review of International Affairs 14(1): 277–294).

Craig, J. The piligrim's progress, Newsweek, 31 August 2018.

Deloitte. 2015 Global health care outlook: Common goals, competing priorities. Available at: https://bit.ly/1OdWJEz.

Deloitte. National transformation in the Middle East a digital journey. Available at: https://bit.ly/2B3GdJC.

Dubai Future Accelerators. Available at: https://bit.ly/2YmEngp.

Dubai.ae. Dubai health authority and Agfa healthcare chest x-ray AI Algorithm preliminary results show 95 per cent accuracy, 4 November 2018. Available at: https://bit.ly/2IZ1fgz.

Economist Intelligence Unit. Saudi Arabia plans for an AI future, 27 July 2018. Available at: https://bit.ly/2WRPU6w.

Elsaadani, A., Purdy, M., and Hakutagwi, E. 2018. Pivoting with AI, Accenture Consulting.

Emirates247. MOHAP launches first smart app to help patients with Multiple Sclerosis, 30 January 2018. Available at: https://bit.ly/2IWtw7E.

Feldbaum, H., and Michaud, J. 2010. Health diplomacy and the enduring relevance of foreign policy interests. PLoS Medicine 7(4): 1–5.

Fidanci, O. How technology can revolutionize healthcare across the MENA region, World Economic Forum, 04 April 2019. Available at: https://bit.ly/2IjDlMU.

Government of Dubai. DHA signs four MoUs to adopt inventive healthcare innovations at the Dubai Future Accelerators, 18 May 2018. Available at: https://bit.ly/2IP3OPW.

Government of Dubai. Dubai Medical Tourism Industry generated more than AED 1.4bn for the Emirate in 2016, 06 June 2017. Available at: https://bit.ly/2JydEK2.

Government of Dubai. His Highness Sheikh Hamdan bin Rashid Al Maktoum launches Tawreed AI at Arab Health 2019. Available at: https://bit.ly/2K3xpZn.

Gulf News. UAE and India sign agreement on Artificial Intelligence, 28 July 2018. Available at: https://bit.ly/2YRio0G.

Gulf Times. Qatar's health system ranked 5th best globally, 6 March 2019. Available at: https://bit.ly/2w6fRDB.

Health Diplomats. Health diplomacy, Genève. Available at: https://bit.ly/2PS6T61.

HiNT Neuro. Available at: https://www.hintneuro.com/.

Imperial College London Diabetes Centre (ICLDC). Available at: https://bit.ly/2JuD3nT.

INSEAD. The Healthcare Sector in the United Arab Emirates, Innovation Brief No. 4. Available at: https://bit.ly/30yFniT.

Jack, A. Cuba's medical diplomacy, Financial Times, 15 May, 2010.

Jefferson, R. New research finds Artificial Intelligence can predict premature death, Forbes, 29 March 2019.

Kickbuch, I. and Rosskam, E. 2012. Introduction: The art and practice of conducting global health negotiations in the 21st century. In Kickbuch, I. and Rosskam, E. (ed.), Negotiating and navigating global health: Case studies in global health diplomacy. London: World Scientific Publishing.

Kickbusch, I. Kokney, M. 2013. Global health diplomacy: Five years on, Bulletin World Health Organization, 1 March, 91(3): 159–159A.

Kickbusch, I. 2003. Global health governance: Some new theoretical considerations on the new political space. In Lee, K. (ed.), Globalization and health, pp. 192–203. London: Palgrave.

Kickbusch, I. and Behrendt, T. 2017. Oxford bibliographies: Global health diplomacy. Available at: https://www.oxfordbibliograhies.com.

Kickbuscha, I., Silberschmidt, G., and Buss, P. 2007. Global health diplomacy: The need for new perspectives, strategic approaches and skills in global health, Bulletin of the World Health Organization, March 2007, 85(3).

KPMG and World Government Summit. Perspectives on government services, cities and technology. Available at: https://bit.ly/2M8LHub.

KSA Government. 2019. Saudi vision 2030. Available at: https://bit.ly/2Xx5aVU. Kuwait Vision 2035. Available at: https://bit.ly/30RCCrP. Bahrain Economic Vision. https://bit.ly/2Ox423N.

Lee, K. 2009. Understandings of global health governance: The contested landscape. In Kay, A. and Williams, O. (eds.), Global health governance: Crisis, institutions and political economy, pp. 27–41. London: Palgrave.

McInnes, C., and Lee, K. 2006. Health, foreign policy and security. Review of International Studies 32(1): 5–23.

McKinsey and Company. The future of jobs in the Middle East, January 2018. Available at: https://mck.co/2ELF3Hl.

Medical Tourism Index. Available at: https://bit.ly/2YyDCQY.

MENA Research Partners. Report, 29 March 2017. Available at: https://bit.ly/2WWXWuO.

Metz, C. Making new drugs with a dose of Artificial Intelligence, The New York Times, 5 February 2019.

Michaud, J., and Kates, J. Global health diplomacy: Advancing foreign policy and global health interests. Global health: Science and practice, Vol. 1, No. 1, Kaiser Family Foundation, 2013. Available at https://bit.ly/2Lv7GK3.

Ministry of health and prevention. Ministry of Health launches project Tele-ICU for remote critical care, 26 January 2016. Available at: https://bit.ly/2LKRgyY.

Mubadala Healthcare to open Wooridul Spine Centre in Dubai this year, Press Release, 11 January 2011. Available at: https://bit.ly/2Jv9kv9.

Myers, M. Artificial intelligence can predict survival of ovarian cancer patients, UCL, 15 February 2019. Available at: https://bit.ly/2TP2QIY.

Nagel, N. Saudi Vision 2030: The opportunities in healthcare. Available at: https://bit.ly/2w5s0IS.

Naseba. The A- Z of Saudi's NTP technology spending spree. Available at: https://bit.ly/2ZStUcO.

Nye J. 2005. Soft power: The means to success in World politics. Washington DC: Public Affairs.

Patients Beyond Borders. Medical Tourism Statistics & Facts, 25 January 2019. Available at: https://bit.ly/2kUqrL8.

Pereira, N. $61bn GCC healthcare projects in the pipeline for 2018, ConstructionWeekOnline. Available at: https://bit.ly/2VjUEAj.

PWC. The potential impact of Artificial Intelligence in the Middle East, 2018. Available at: https://pwc.to/2C6NQAV.

PWC. Why AI and robotics will define New Health, June 2017. Available at: https://pwc.to/2oSziNv.

Qatar Computing Research Institute. National Artificial Intelligence Strategy for Qatar, Hamad bin Khalifa University. Available at: https://bit.ly/2Tt2qLs.

Qatar Genome. Qatar Genome programme joins forces with Genomics England, 17 April 2019. Available at: https://bit.ly/2JLttNw.

Qatar Government. 2019. Qatar National Vision 2030. Available at: https://bit.ly/35hQxLq.

Reachingthelastmile.com. 2019. Reaching the Last Mile. Available at: https://bit.ly/2VGiag6.

Roberts, E. Qatar offers thousands of expat jobs at pioneering medical Centre, The Telegraph, 21 May 2014. Available at: https://urlzs.com/wUZrY.

Rose, D. New university in Abu Dhabi focuses on artificial intelligence, The Times, 25 March 2020. Available at: https://bit.ly/3gGia5a.

Salama, S. UAE looking for a technological leap, Gulf News, 21 October 2017. Available at: https://bit.ly/2YFAjI6.

Salama, S. COVID-19: Saudi Arabia ranked 3rd globally in using technology to fight coronavirus, Gulf News, 18 June 2020. Available at: https://bit.ly/2EHifZx.

Saudi Gazette. NDU and GE agree to drive digital industrial innovation, 16 November 2017. Available at: https://bit.ly/2imzvoz.

Sharif, A. Saudi Puts Money on Tech as It Prepares for Life After Oil, Bloomberg, 13 August 2018. Available at: https://bloom.bg/2VV2jJz.

Stuckler, D., and McKee, M. 2008. Five metaphors about global-health policy. Lancet 372: 95–97.

The Arab Hospital Magazine. Artificial Intelligence in medicine. Available at: https://bit.ly/2WxaD2G.

The Economist Intelligence Unit. Artificial Intelligence in the Real World: The business case takes shape, A Report from the Economist Intelligence Unit Limited, 2016.

The Economist Intelligence Unit. Saudi Arabia plans for an AI future.

The National Digitization Unit (NDU). Saudi Arabia. Available at: https://bit.ly/2WPN9Tf.

The U.S.-U.A.E. Business Council. January 2018. https://bit.ly/2LYauRW.

UAE Government. 2019. UAE Vision 2021. Available at: https://bit.ly/31TNlUc.

UAEBARQ. DHA signs four MoUs to adopt AI healthcare innovations, 17 May 2018. Available at: https://bit.ly/30Ns8eq.

UN General Assembly. Global health and foreign policy: Strategic opportunities and challenges, note by the Secretary General. 23 September, 2009, A/64/36. Available at: https://bit.ly/2QC6Dsq.

UNCTAD. 2014. Transfer of technology and knowledge sharing for development, current studies on science, technology and innovation, N. 8., UN Publications.

United Nations. 2009. Resolution adopted by the UN General Assembly on "Global Health and Foreign Policy" (A/Res/64/108), Sixty-forth Session.

WISH Data Science and AI Forum. 2018. Report of the WISH data science and AI forum, harnessing data science and AI in healthcare from policy to practice. Available at: https://bit.ly/2QsGoUW.

World Health Organization. 2018. Public spending on health: A closer look at global trends, Global Report. Available at: https://bit.ly/2TWkPgI.

World Health Organization. Global health diplomacy. Available at https://bit.ly/2XAmHkj.

CHAPTER 8

Free Zones in Dubai: Accelerators for Artificial Intelligence in the Gulf

Robert Mogielnicki

1 Introduction

There exists substantial excitement surrounding the economic potential of artificial intelligence (AI) in Gulf Arab states; however, the importance of local knowledge, rigid commercial policies, and social risks present barriers to the effective implementation of AI across the region's economies. Free zones represent commercial spheres of exception wherein business actors can circumvent strict commercial policies in the onshore economy. For this reason, academic literature categorizes free zones as litmus tests for reforms in the broader economy, catalysts of change in specific sectors, and mechanisms for transferring technology and skills to local populations (Moberg 2017; Lu 2014; Johansson and Nilsson 1997). Therefore, free zones offer—in both theory and practice—a seemingly hospitable ecosystem for AI technologies. Perhaps more importantly, free zones present a controlled environment for the government to encourage

R. Mogielnicki (✉)
The Arab Gulf States Institute, Washington, DC, USA
e-mail: robert.mogielnicki@agsiw.org

E. Azar and A. N. Haddad (eds.), *Artificial Intelligence in the Gulf*,
https://doi.org/10.1007/978-981-16-0771-4_8

greater utilization of AI services and applications on the part of the private sector.

This chapter will assess the degree to which free zones can serve as useful vehicles for the effective implementation of AI technologies in Gulf Arab economies by exploring the case study of Dubai. The emirate of Dubai contains around 24 operating free zones, which account for an estimated 354,000 jobs and 41 per cent of the emirate's total trade. Many of the emirate's free zones are dedicated to technological innovation: Dubai Silicon Oasis promotes tech-focused entrepreneurship, CommerCity is the Gulf region's first e-commerce free zone, and Dubai South, a state-owned enterprise, pledged $545 million to develop the emirate's second e-commerce free zone. The 50-year charter announced by Dubai's ruler demonstrates the emirate's long-term commitment to a free zone-led growth model that is heavily reliant on the creation of a knowledge-based economy. The charter's articles include plans to develop a virtual commercial zone capable of servicing 100,000 firms, and these initiatives broadly align with the Smart Dubai 2021 strategy to make the emirate the happiest city on earth by embracing technological innovation.

The Dubai government's strategy to embrace technological innovation requires participation by private sector actors. However, large foreign firms operating in Dubai and the broader UAE usually opt to establish headquarters for their research and development departments in American, European, or Asian cities. Moreover, private sector firms across Gulf Arab economies have traditionally viewed public sector spending as the main driver of growth in the region. As this view changes, the resulting commercial environment in cities like Dubai will play a key role in determining the supply and demand dynamics associated with AI platforms and services.

By combining the topics of AI innovation strategy and free zone development in the Gulf region, this study represents the first academic investigation of its kind. The work follows a multi-disciplinary, political economy approach—a useful analytical lens for mapping and assessing the implementation of AI concepts and technologies across free zones. The resulting chapter outlines the government-led emphasis on AI in Dubai and, to a lesser extent, the rest of the UAE and then maps the existing technology-focused free zone infrastructure in Dubai. The chapter next explores how the proliferation of AI technologies may simultaneously shape the emirate's free zone system in the future, with a focus on the development of talent pipelines in educational institutions and promoting public-private partnerships in the finance, trade and logistics, and retail sectors. The work also addresses the social and political ramifications of efforts to better develop and deploy AI technologies within free zones, given that Dubai's free zones constitute a critical source of government revenues and employment opportunities. The conclusion includes policy recommendations for better aligning AI innovation with free zone development.

2 Artificial Intelligence in Dubai: A Government Priority

Gulf Arab states seek to develop tech-driven, knowledge economies; advanced technologies such as AI and blockchain fit well with economic trajectories that emphasize new contributions from high-impact sectors. The UAE—and in particular the emirate of Dubai—has identified AI as a central component of its tech-driven economic strategy, and the government hopes to position the country as a "global incubator" (Government of the UAE 2019) for commercial projects and ventures that implement AI-focused technologies. Underlying government-level interest is dependent upon AI's expected contribution to the global economy. According to recent studies, AI could add $15.7 trillion dollars to global gross domestic product by 2030 (PwC 2017) and enable the creation of between $3.5 trillion and $5.8 trillion in value annually—equating to between 1 and 9 per cent of revenue in 2016 across various industries (McKinsey 2018).

Multiple government entities across the UAE are charged with overseeing the implementation of AI in government processes as well as with creating linkages to the private sector. The Minister of State for Artificial Intelligence, Omar Al Olama, who was appointed in October 2017, manages the country's AI portfolio. The Ministry of Cabinet Affairs and the Future, which handles the execution of all internal and external affairs related to the country's federation, also works closely on AI initiatives. The work of these ministries falls under the UAE Artificial Intelligence Strategy 2031, which aims to boost the country's gross domestic product by 35 per cent, reduce government costs by 50 per cent, and make the country 90 per cent resistant to a financial crisis. While the final goal of the AI strategy is difficult to measure because it relies on a counterfactual scenario, the objective nevertheless demonstrates the degree to which AI technologies are embedded in the UAE government's economic strategies. Indeed, these efforts are already paying dividends: the UAE ranked within the top 20 countries globally and higher than China in a Government AI Readiness Index produced by Oxford Insights (Miller and Stirling 2019), and Dubai and Abu Dhabi ranked sixth and tenth in terms of the most appealing cities for global digital experts (Trade Arabia "2019).

Dubai, perhaps more than any other Gulf Arab polity, views AI as a lynchpin of its drive to incorporate advanced technologies within government processes and develop a highly-skilled private sector. The Dubai Future Foundation and Smart Dubai, a government body tasked with driving the emirate's "smart" technological transformation, represent two government entities involved in promoting AI technologies in the emirate. Smart Dubai runs an AI Lab in partnership with IBM to provide the tools and training for government and private sector actors to implement AI services and applications within their respective fields. This promotion of AI manifests in emirate-level strategies, such as the Dubai Autonomous Transportation Strategy. The strategy aims to transform 25 per cent of the emirate's transportation to autonomous modes by 2030; such a transformation would involve around five million daily trips and potentially save ~ $6 billion per year in transportation costs (Dubai Future Foundation 2016).

Interest in AI extends beyond broad strategies and ministerial portfolios. Dubai's National Program for Artificial Intelligence hosted a major conference, AI Everything, in the spring of 2019 under the patronage of Sheikh Mohammed bin Rashid. Moreover, the Dubai Police are developing fully-automated police stations and have arrested 319 criminal suspects using AI-powered CCTV devices as of March 2019 (Debusmann 2019). The rapid pace of AI technology adoption—both conceptually within Dubai's economic strategies and in actual government processes—increased the need for clear policy guidelines and frameworks. The Smart Dubai AI Ethics Principles and Guidelines document offers an initial step towards addressing these needs. The Dubai government likewise mandated the Dubai Public Policy Research Centre to study how the introduction of AI technologies will affect citizens' perceptions of identity and the role of government.

Despite the high-level focus on AI and subsequent excitement surround associated technological applications, Dubai's policymakers understand that implementing AI involves several stages. The Dubai government, for example, aims to digitize all government processes before the launch of Expo 2020. Although digitization represents a rudimentary step from a technological standpoint, it is necessary for implementing AI on a larger scale across the economy. Unlike the rest of the globe, wherein the private sector operates as the primary investor in AI technologies, the public sector plays an outsized role, accounting for as much as 20 per cent of all investments and second only to the finance sector, in the Middle East and Africa (International Data Corporation 2019). Public sector spending on AI, therefore, has the potential to conflict with broader efforts to spur the private sector to play a larger role in Gulf economies like Dubai, where the private sector has relied on the government to be the main investor in AI. Free zones offer controlled, commercial spheres wherein public and private firms may collaborate to expand the availability of AI services and applications and share the burden of investments.

3 Dubai's Free Zone System: A Long-Term Strategy

Despite the current hype surrounding AI and other innovative technologies in Dubai, the emirate's commercial reputation rests in large part on the foundations created by its free zone system. Indeed, Easa Saleh Al-Gurg, a prominent UAE government advisor and former long-standing ambassador to the United Kingdom, described the Jebel Ali port and free zone as having "contributed more, perhaps, than any other innovation to the present success which Dubai enjoys" (Al-Gurg 1998). The Jebel Ali Free Zone, considered the Gulf region's first, emerged in Dubai during the early 1980s. When the Jebel Ali Free Zone officially opened in 1985, it hosted 19 companies. By 2016, it sustained more than 7,000 fee-paying companies, 207,000 jobs, and 60,000 residents. Moreover, Jebel Ali Free Zone attracted approximately 32 per cent of the UAE's foreign direct investment in 2015 (JAFZA 2015).

As the central nodes of commercial activity for foreign investors, free zones play a central role in Dubai's political economy. Free zones can be broadly defined as demarcated geographic areas contained within a territory's national boundaries where the rules of business are different from those that prevail in the national territory (Farole 2011, 23). The relative economic autonomy of individual emirates within the UAE, an integral country case when examining free zones in the Gulf region, suggests that free zones are better conceptualized within autonomous territories rather than within national boundaries by default. More specifically, free zones in the Gulf differ from their onshore counterparts in three crucial areas: permitting full foreign ownership of commercial entities, offering reduced workforce nationalization requirements, and providing duty and tax exemptions. The free zones observed in this work demonstrated at least two of the three aforementioned characteristics. It is important to note that other economic zones in the Gulf, like industrial zones and industrial estates, do not differ substantially from the rules regulating onshore commercial activity in their territories and therefore are not considered free zones in this analysis.

There existed 24 operating free zones in the emirate by the end of 2016.[1] Dubai's broader free zone system managed between 90 and 91 per cent of the UAE's total free zone trade during 2014 and 2015 (Federal Customs Authority 2015), and free zones in Dubai contributed approximately 30 per cent to the UAE's total trade in 2015. Free zones play a central role in economic diversification—these commercial entities facilitated 41 per cent of the country's non-oil trade in 2018. One of the newer additions to Dubai's free zone sector is CommerCity, a $735 million e-commerce-focused venture. Dubai Airport Freezone and *Wasl* Asset Management Group, a major real estate company created by the Dubai Real Estate Company and chaired by Sheikh Maktoum, jointly manage the e-commerce free zone (Dubai Government Media Office 2017).

The free zone sector in Dubai is poised for further growth; the emirate's 50-year charter illustrates Dubai's steadfast, long-term commitment to a free zone-led growth model. On 4 January 2019, the ruler of Dubai released his vision for the future of Dubai with a charter containing nine articles and stretching over the next five decades. The significance of this charter can be gleaned from the level of elite involvement: the ruler will "personally observe" progress on these articles and has tasked Dubai's crown prince, Sheikh Hamdan bin Mohammed bin Rashid Al Maktoum, with the duty of overseeing the charter (Mohammed bin Rashid Al Maktoum 2019). The document is broad in scope and light on implementation details—a pervasive feature of country visions, strategies, and other agendas in Gulf Arab states.

The charter frames the socioeconomic development of Dubai's citizens through a prism that emphasizes the role of free zones and other forms of special economic zones. Often referred to as "free zones" in Dubai,

[1] Figures based on author's fieldwork and site visits in Dubai.

special economic zones are "geographically delineated areas subject to differentiated regulation and administration from the host country in which it resides, for the purpose of attracting foreign direct investment in economic activities that could not otherwise be achieved" (Oliver Wyman 2018). In the second article of the charter, Dubai's ruler proposes the creation of a geo-economic map within the emirate as a precursor for expanding the proliferation of special economic zones. Beyond examples of physical zones, Dubai policymakers want to further develop the emirate's digital commercial infrastructure. Article three aims to establish a virtual commercial zone capable of hosting 100,000 companies, and part of this program involves streamlined banking and immigration processes—two service areas wherein bureaucracy tends to limit the ease of doing business.

The creation of e-commerce and virtual free zones combined with the promotion of AI across government and private sector processes enhances Dubai's ability to strengthen partnerships with key global actors involved in AI research and development. The virtual commercial city concept being promoted in Dubai resembles the Malaysian Digital Free Trade Zone, a joint venture between the government of Malaysia and Alibaba, the Chinese e-commerce firm owned by billionaire Jack Ma. Malaysia's digital free zone aims to handle $65 billion worth of goods and create 60,000 jobs through an electronic platform designed to ease trade between Chinese and Malaysian companies. The Digital Free Trade Zone is likewise considered an official Belt and Road Initiative project (Chandra 2018). In replicating an Asian model for a Gulf Cooperation Council (GCC)-based virtual free zone, Dubai seeks to position itself within the *digital* Belt and Road Initiative. This manoeuvring will attract greater interest from China, a global powerhouse in AI research and patents, as well as Chinese firms. For example, Huawei possesses an AI strategy that involves investing in AI research and engaging globally with academia, industries and partners to develop and an AI ecosystem and talent pipeline.

Articles four and six of Dubai's 50-year charter focus on commercial outcomes to educational attainment by utilizing free zones. The charter notes, "We aim to build an educational and learning system that explores and develops people's skills" (Mohammed bin Rashid Al Maktoum 2019), and a central component of this system involves transforming public and private universities into free zones in order to "allow students to carry out business and creative activities, make these activities part of the education and graduation system, and shape integrated economic and creative zones around the universities" (Ibid). The objective of this educational strategy is to move beyond graduating students and to generate new companies and employers. This offers an opportunity to increase the availability of course offerings on AI in the emirate, which are growing but remain limited. The British University of Dubai launched the first bachelor's degree in AI in 2018, and the United Arab Emirates University has a minor in AI as part of its course offerings.

The commercial and academic focus on AI represents a nexus between Dubai's development aspirations and China's technological research and innovation capabilities. Chinese universities play a dominating role in the ability to produce inventions related to distributed AI, machine learning techniques, and neuroscience/neurorobotics (WIPO 2019). Students from China are well-positioned to support Dubai's AI initiatives: there were approximately 30,000 Chinese students in Dubai as of 2019, and several educational institutions across the emirate reported a twenty per cent year-on-year increase in Chinese undergraduates deciding to study in Dubai (Rivzi 2019). However, Dubai must convince Chinese students of the academic and commercial benefits of pursuing AI studies in the UAE, as opposed to studying in China, or otherwise develop strong partnerships with Chinese universities and research institutes focusing on AI.

4 A Public-Private Partnership for AI?

The alignment of AI initiatives and free zones in Dubai involves more than just overlapping development strategies. The structure of Dubai's free zones, wherein government-funded zones oversee the commercial activities of predominantly private sector firms, enables the government to encourage greater employment of AI technologies within the emirate's private sector. According to International Data Corporation figures on global AI spending patterns, the Middle East and Africa lags far behind the rest of the world in AI spending. Total global spending on AI reached $24.86 billion in 2018, whereas the Middle East and Africa spent a mere $37.49 million in 2017 (International Data Corporation 2019). The public sector in the Middle East and Africa played an outsized role in AI spending—second only to the finance sector—and therefore heavily influenced the top expected uses for AI technologies, which included defence, intelligence, and automated customer service. Across the rest of the globe, the private sector drove spending on AI. The resulting top uses for AI included commercially-focused implementations: customer service, sales processes, and threat intelligence and prevention. As Gulf Arab countries seek to reduce public sector expenditures, governments will need to transfer much of the responsibility for AI research, development, and implementation to the private sector. Free zones provide a convenient mechanism for channelling the Dubai government's ambitions for AI while also promoting these technologies within the emirate's private sector.

Financial free zones, such as the Dubai International Financial Centre (DIFC), offer an obvious starting point for the adoption of AI technologies. The DIFC's independent legal framework, which is based on English common-law, enhances the free zone's structural isolation, but it remains embedded as a central commercial and urban structure of the emirate. At the same time, the DIFC works closely with the country's central bank, thus permitting an institutional mechanism by which AI-specific knowledge and practices can be shared across the entire industry. With respect to AI investments, the finance sector outspent all other sectors in the Middle East and Africa, reflecting strong private sector interest. In addition to strong demand, the potential economic pie is large. It is estimated that AI can save the global financial institutions approximately $1 trillion by 2030 in front office, middle office, and bank office activities (Autonomous Research 2019). Chatbots, anti-fraud and risk assessments, and credit underwriting are currently employed within the sector. Moreover, voice assistants, monitoring, and smart contracts infrastructure are increasingly being utilized by financial institutions.

The trade and logistics sectors represent an important nexus of the country's economic diversification efforts, free zone activity, and potential for adoption of AI technology. More than one-quarter of the UAE's entire trade is handled by free zones, and the country possesses the most advanced transport infrastructure in the Gulf (Jensen 2018). "In logistics, the network-based nature of the industry provides a natural framework for implementing and scaling AI and amplifies the human components of highly-organized global supply chains", finds a separate report by DHL and IBM (DHL & IBM 2018). The ability to better optimize trade and logistics networks and better orchestrate digital supply chains would be a major advantage for firms operating in free zones that cater to these sectors. Leveraging cognitive automation—generally understood as the combination of robotic process automation and AI—to manage financial anomalies, customs, contracts, and customer information would help free zone firms better operate global supply chains. The Jebel Ali Free Zone, Dubai Airport Freezone, Dubai Multi Commodities Centre (DMCC), and the Logistics District of Dubai South are prominent examples of free zones engaged in trade and logistics in Dubai. The Jebel Ali Free Zone and DMCC already share a customs corridor that permits free zone firms to utilize the infrastructure of both entities and avoid onshore regulations; this collaboration could be further enhanced by shared AI platforms. Dubai Customs also signed a memorandum of understanding with the Dubai Free Zones Council to implement a "virtual stock guarantee" initiative that would support re-export activity from free zones to external markets (Logistics Middle East 2019)—another initiative ripe for AI applications.

Free zones such as CommerCity, Dubai Design District (d3), and the Gold and Diamond Park serve as excellent environments to experiment with AI's ability to transform retail and e-commerce experiences. For physical retail locations, such as d3 and Gold and Diamond Park, AI can help retailers personalize in-store experiences for customers and maximize the revenue for a given unit of floor or shelf space. In other free zones that prioritize online shopping experiences and e-commerce, such as CommerCity, AI can be employed to better identify targeted customer segments and provide relevant recommendations for products and services to a wider audience (DHL & IBM 2018). With many of the clients in these free zones representing small and medium enterprises or regional branches of larger firms, free zone authorities will likely need to make the initial investments in AI services and grant access to participating firms.

Recent changes to the economic institutions surrounding free zone activity in Dubai and the wider UAE create an urgency for new technologies that permit free zones to retain their comparative advantage, on the one hand, and increase the likelihood of technology spillovers both across free zones and into the onshore economy. In late 2018, the UAE passed a new foreign ownership law allowing foreigners the right to maintain full ownership of commercial ventures in select sectors of the economy (Anderson 2019). Since one of the main commercial advantages of free zones in the UAE involves the right to 100 per cent foreign ownership, the changes to ownership regulations across the country partially reduced the commercial attractiveness of Dubai's free zones. Free zones across the country cut fees for rent and services in order to retain their customers. Rather than cost-cutting, free zones should seek to add value by better incorporating new AI technologies throughout the emirate. Investments in AI technology could be shared across the emirate's free zones as new initiatives pick up steam. For example, the Dubai Free Zone Council reached a preliminary agreement wherein any free zone firm in Dubai can access all of the emirate's free zones without needing additional licences.

Greater integration between free zones will reduce overall costs associated with acquiring AI technology services and applications and also maximizing its impact on free zone firms. Over the past couple of years, free zones like Dubai Airport Freezone have introduced dual licences that permit firms to access both onshore and offshore markets, creating a more flexible investment procedure. Dual licences expand the scope of technological spillover beyond the offshore economy and into the onshore economy.

5 Assessing Broader Implications

While changes to free zone laws and regulations present opportunities for enhancing the access to and impact of AI technologies, free zones must ensure that their client base generates sufficient demand for AI platforms and services. Reorienting the composition of free zone clients and future marketing efforts towards geographic locations with a track record of spending on AI is a necessary step in this regard. According to estimates by the International Data Corporation, the United States will be responsible for 60 per cent of the global spending on AI, with Western Europe and China being the next largest spenders on cognitive and artificial intelligence systems (International Data Corporation 2019). Dubai's free zones, however, tend to attract a disproportionate number of firms from territories with historically low levels of investment in AI. Historical ties with the Indian subcontinent and weak trade linkages amongst GCC territories likewise influenced the nature of capital investments in Dubai's free zones.

The breakdown of local, regional, and global clients in the Dubai Technology Entrepreneurship Centre, a program of Dubai Silicon Oasis free zone, offers a useful visualization of capital flows through free zones. Moreover, this technology-focused centre provides an excellent case study for examining how the composition of clientele may influence AI technology adoption in the future. Indian companies represented the largest national segment of clientele, occupying 30 per cent of the free zone's clients, and Pakistani firms contributed another 7.3 per cent of the total.[2] Alternatively, GCC companies accounted for a mere five per cent of clients. Non-GCC, Middle Eastern firms provided the second-largest segment of clients after Asian firms. Much of this investment originated in war-torn or struggling economies like Syria, Egypt, and Jordan, which together constituted approximately 78 per cent of Middle Eastern clients. Therefore, more work could be done within Dubai's free zones to attract clientele from countries with proven records of spending on AI.

The relatively small number of local Emirati firms operating within Dubai Technology Entrepreneurship Centre suggests that greater adoption of AI technologies within free zones would not substantially impact local firms, thus minimizing the in-country value of this investment. Only 23 Emirati companies operated amongst the free zone's total client base of 802 firms. Free zones permit firms to circumvent national quotas of Emirati employees—dictated by a broad labour market policy known as Emiratisation—and consequently, very few Emiratis function as employees in free zone firms. As most free zones are partially or fully government-owned, their workforces tend to contain higher ratios of Emirati citizens as a portion of the total workforce. Therefore, the nationally-oriented processes of skills transfer and upskilling associated with AI technologies are more likely to occur within the bureaucracies of free zones rather than on a meso-level amongst Emirati-owned firms operating within free zones.

The Dubai Free Zone Council functions as an institutional coordination mechanism within the emirate's free zone system. This council can aid in coordinating investments in AI platforms and their subsequent implementation across the emirate's free zones. Indeed, other coordinating efforts associated with dual offshore/onshore licensing and permitting free zone firms to operate across Dubai's free zones without holding multiple licences can further facilitate the dissemination and impact of AI technology beyond the clearly demarcated boundaries between offshore free zones and onshore businesses. Sharing AI innovations with UAE free zones located outside of Dubai poses a greater challenge.

[2]The following figures refer to registration numbers as of December 2017. The author remotely collected this data on 13 December 2017 from an employee of Dubai Technology Entrepreneurship Centre, a program of Dubai Silicon Oasis.

UAE-wide interest in AI exists: Abu Dhabi Global Market launched Hub 71 to enable the innovation and growth of transformational technology companies, and the Department of Ports and Customs in Ajman is working with a boutique consulting firm to incorporate AI into trade practices and record keeping.[3] However, the UAE's national government lacks an institutional mechanism for coordinating free zone activity across emirates—this is symptomatic of the autonomy over economic affairs afforded to emirate-level governments by the constitution. Competition between the country's federal entities reduces the incentives associated with knowledge-sharing. The country's central bank represents one of the few exceptions because it does have the power to influence the behaviour of financial free zones, such as Dubai International Financial Centre and Abu Dhabi Global Market.

Exploiting the homogeneity of the country's free zones, especially those clustered within specific sectors and industries, remains a promising approach for sharing AI knowledge and innovations across the country and addressing the problem of incentives. Here, entities like the World Free Zones Organization can play an important role in building bridges between sector-specific clusters of free zones and reducing competitive commercial behaviour.

Historically, commercial cooperation between UAE free zones operating in similar sectors of the economy has been limited. The Ras Al Khaimah Media Free Zone, which opened in 2006 with the strategy of functioning as a cheaper version of Dubai Media City, closed following the launch of additional media-focused free zones in Abu Dhabi and Fujairah. Industry-specific AI services and applications, though, present an opportunity for greater collaboration between government-run free zones on the UAE's Artificial Intelligence Strategy 2031, which calls for "investment in the latest AI technologies and tools to enhance government performance and efficiency" (Government of the United Arab Emirates 2019) (Table 1).

Other GCC countries may possess broader digital strategies, like Qatar's Digital Government Strategy 2020 or specific components of Saudi Arabia's National Transformation Plan 2020, but do not possess established and coherent AI strategies. Saudi Arabia did, however, launch

[3] Personal interview, partner at a consulting firm, Dubai, United Arab Emirates, May 9, 2018.

Table 1 Sector breakdown of UAE free zones; the list is not comprehensive and may include free zones under construction

Free Zone Sectors	*Dubai*	*Abu Dhabi*	*Northern Emirates*
Finance	Dubai International Financial Centre	Abu Dhabi Global Market	N/A
Aviation	Dubai Airport Freezone; Dubai South	Abu Dhabi Airports Free Zone	Sharjah Airport International Free Zone
Maritime	Jebel Ali Free Zone	Khalifa Industrial Zone Abu Dhabi	Ajman Free Zone; Fujairah Free Zone; Ras Al Khaimah Maritime Free Zone
Media	Dubai Media City	Twofour54	Creative City; Sharjah Media City Free Zone; Ajman Media City Free Zone
Healthcare	Dubai Healthcare City	N/A	Sharjah Healthcare City
Technology	Dubai Internet City; Dubai Silicon Oasis; Dubai Science Park; Commer City	Masdar City	Sharjah Research Technology and Innovation Park

a national strategy for data and AI in October 2020. Moreover, the structural aspects of Gulf free zone systems outside of the UAE hamper the horizontal and vertical integration and sharing of AI services and applications. Non-UAE free zone systems tend to be smaller—both in terms of the number of entities and overall contribution to the economy—and, despite the lack of federal political system, competitive dynamics persist in the form of intergovernmental competition and competing interests of the public and private sectors. This leaves the potential for tech-focused enclaves, such as Knowledge Oasis Muscat, to carve out niche programs for AI. Alternatively, international business-to-business partnerships could foster cross-border spillovers in AI innovation. The Saudi Economic Cities Authority, for example, signed a memorandum of understanding with the Dubai Airport Freezone in October 2017 (*Al Khaleej* 2017). However, the GCC-crisis that emerged in June 2017 made it extremely difficult to pursue similar regional collaborations with Qatari free zones. Meanwhile, Omani economic policymakers have expressed concerns about commercial collaboration with the UAE.[4]

6 Conclusion

Dubai's recent focus on AI innovation as part of a country-wide strategy to transform the UAE into a technology-driven, knowledge economy should not be viewed in isolation from long-standing economic strategies based on free zone-led development. In fact, structural factors associated with Dubai's free zone sub-system and the nature of its engagement with other free zone sub-systems in the country can help to address talent pipeline, investment, knowledge sharing, and cooperation issues related to AI. Dubai's 50-year charter likewise reaffirms the emirate's long-term commitment to free zone development, which provides additional justification for better aligning these two economic development strategies. The following are specific opportunities that can boost and sustain such alignment:

1. *Talent Pipeline*—Dubai International Academic City free zone contains 27,000 students and over 500 academic programs, and the Dubai government aims to transform all of the public and private universities into free zones. This educational transformation, wherein educational attainment is directly linked to commercial outcomes, offers an opportunity to expand course offerings on AI—thereby developing an indigenous talent pipeline—and to market academic programs to students from AI research hubs, such as China.
2. *Public-Private Partnerships*—As Gulf Arab countries continue to make strides in reducing public sector expenditures, governments must transfer much of the responsibility for AI research, development, and implementation to the private sector. Free zones provide a convenient mechanism for channelling the Dubai government's ambitions for AI while also promoting related services and applications within the emirate's private sector. Dubai free zones are

[4] Leaked government cables suggested that Oman's rebuff of a UAE offer to purchase a 50 per cent stake in the Salalah Free Zone stemmed from fears that the Emiratis would dump undesirable free zone clients into the zone.

predominantly state-owned enterprises, but the majority of clients operating within free zones are private sector firms. Thus, investments in AI services and applications within free zones impact both the public and private sectors. The increasing employment of dual licences within free zones also increases the likelihood of technology spillovers from offshore markets and onto onshore markets.

3. *Knowledge Sharing*—The scale of Dubai's free zone sector, with more than two dozen free zones, and nearly twenty more free zones scattered across Abu Dhabi and the northern emirates means that several of the country's free zones share a sectoral focus. Exploiting the regulatory homogeneity of the country's free zones, especially those clustered within specific industries, remains a promising approach for sharing AI knowledge and innovations across the country in an efficient and cost-effective manner. Free zones engaged in finance, trade and logistics, retail and e-commerce, healthcare, and technology are particularly well-positioned to capture benefits from AI innovation.
4. *Fostering Cooperation*—Sharing AI services and applications between free zones requires mitigating the competitive dynamics within and between emirates. The Dubai Free Zone Council can promote collaboration as it pertains to Dubai's economic affairs, and the UAE Central Bank possesses some oversight over financial free zones. Organizations like the World Free Zone Organization can promote dialogue but have no authority over economic policy. However, a national free zone committee connected to the UAE Artificial Intelligence Strategy 2031 would help foster greater collaboration over AI services and applications that impact free zone activities.

The Dubai government's efforts to promote AI within the public sector are commendable; however, AI innovation must occur alongside genuine private sector development. Thus far, the economic strategies pertaining to AI and free zones, the former being a relatively new conceptualization of the UAE's economic diversification efforts and the latter representing a long-standing development approach, have run in parallel. This does not have to be the case. Free zones offer spheres of experimentation for AI services and technology, fertile grounds to cultivate an indigenous talent pipeline, existing models of public-private partnerships, mechanisms for knowledge sharing, and opportunities for country-wide collaboration.

References

Anderson, Robert. "UAE law allowing 100% foreign ownership now in force." Gulf Business. November 5, 2019. https://gulfbusiness.com/uae-100-foreign-ownership-law-now-force/.

Autonomous Research. Augmented Finance & Machine Intelligence. Accessed May 28, 2019. https://next.autonomous.com/augmented-finance-machine-intelligence.

Chandra, Nyshka. "Alibaba's 'Digital Free Trade Zone' has some worried about China links to Malaysia." CNBC. February 12, 2018. https://www.cnbc.com/2018/02/12/concerns-over-alibaba-led-digital-free-trade-zone-in-malaysia.html.

"DAFZA signs MoU with Saudi Economic Cities Authority." Al Khaleej. October 15, 2017.

Debusmann, Bernd. "Dubai Police arrest over 300 suspects with AI-powered cameras." Arabian Business. March 19, 2019. https://www.arabianbusiness.com/technology/415735-dubai-police-arrest-over-300-suspects-with-ai-powered-cameras.

DHL & IBM. Artificial Intelligence in Logistics. DHL Customer Solutions & Innovation. Published 2018. https://www.logistics.dhl/content/dam/dhl/global/core/documents/pdf/glo-artificial-intelligence-in-logistics-trend-report.pdf.

"Dubai Customs signs MoU with Dubai Free Zones Council for Virtual Stock Guarantee." Logistics Middle East. March 18, 2019. https://www.logisticsmiddleeast.com/business/32318-dubai-customs-signs-mou-with-dubai-free-zones-council-for-virtual-stock-guarantee.

Dubai Future Foundation. "Dubai's Autonomous Transportation Strategy." Government of Dubai. https://www.dubaifuture.gov.ae/our-initiatives/dubais-autonomous-transportation-strategy/.

Dubai Government Media Office. "AED 2.7 billion joint venture between DAFZA & wasl to boost Dubai's leading position in global trade." Government of Dubai. October 28, 2017, accessed January 13, 2018. http://www.mediaoffice.ae/en/media-center/news/28/10/2017/dubai-commercity.aspx?TSPD_101_R0=bd6a4e135dc25f75faca5713e7a1c0c2dH10000000000000000c7938c45ffff00000000000000000000000000005a7c5f3d000d36150d.

Farole, Thomas. Special Economic Zones in Africa: Comparing Performance and Learning from Global Experiences (Washington DC: The World Bank, 2011).

Federal Customs Authority. Emirates Customs: Free Zones. Issue 40 (Abu Dhabi: Electronic Statistical Bulletin, 2015).

Government of the United Arab Emirates. Instagram. May 21, 2019. https://www.instagram.com/p/BxuJkSIDcbf/?utm_source=ig_embed.

Government of the United Arab Emirates. "UAE Artificial Intelligence Strategy 2031." Accessed May 30, 2019. http://www.uaeai.ae/en/.

International Data Corporation. "Worldwide Spending on Artificial Intelligence Systems…" Accessed May 21, 2019. https://www.idc.com/getdoc.jsp?containerId=prUS44911419.

Jebel Ali Free Zone. Jafza, Dubai. Where business comes together (Dubai, JAFZA, 2015).

Jensen, Sterling. "Policy Implications of the UAE's Economic Diversification Strategy: Prioritizing National Objectives." In A. Mishrif and Y. Al Balushi. Economic Diversification in the Gulf Region, Volume II: Comparing Global Challenges (Singapore: Springer Singapore, 2018).

Johansson, Helena and Lars Nilsson. "Export Processing Zones as Catalysts." World Development 25.12 (1997).

Lu, Xia. "The Development Process, Function Evaluation, and Implications of World Free Trade Zones." World Review of Political Economy 5.3 (2014).

McKinsey. Notes from the AI frontier: Applications and the value of deep learning. April 2018. https://www.mckinsey.com/featured-insights/artificial-intelligence/notes-from-the-ai-frontier-applications-and-value-of-deep-learning.

Miller, Hannah and Richard Stirling. Government Artificial Intelligence Readiness Index 2019 (London: Oxford Insights, 2019): p. 32. https://ai4d.ai/wp-content/uploads/2019/05/ai-gov-readiness-report_v08.pdf.

Moberg, Lotta. The Political Economy of Special Economic Zones: Concentrating Economic Development (Abingdon: Routledge, 2017).

Mohammed bin Rashid Al Maktoum, "The Fifty-Year Charter," The Executive Office of Dubai, 2019. https://www.mbrmajlis.ae/50-en/The%20Fifty-Year%20Charter.pdf.

Oliver Wyman. Special Economic Zones as a Tool for Economic Development. Published in 2018. https://www.oliverwyman.com/content/dam/oliver-wyman/ME/publications/Special-Economic-Zones.pdf.

PwC, Sizing the Prize, PwC (2017), accessed May 20, 2019. https://www.pwc.com/gx/en/issues/data-and-analytics/publications/artificial-intelligence-study.html.

Rizvi, Anam. "Rising numbers of Chinese students heading for Dubai." The National. April 17, 2019. https://www.thenational.ae/uae/rising-numbers-of-chinese-students-heading-for-dubai-1.850443.

Saleh Al-Gurg, Easa. The Wells of Memory: An Autobiography (London: Albemarle Street, 1998).

Sheikh Mohammed bin Rashid Al Maktoum. "The Fifty-Year Charter." Executive Office of Dubai. January 4, 2019.

"UAE cities among top 10 appealing for digital experts." Trade Arabia. May 30, 2019, http://www.tradearabia.com/news/REAL_355154.html.

WIPO. WIPO Technology Trends 2019: Artificial Intelligence. Geneva: World Intellectual Property Organization. Published 2019.

PART IV

Society, Utopia and Dystopia

CHAPTER 9

AI & Well-Being: Can AI Make You Happy in the City

Ali al-Azzawi

1 Introduction

Ever since Bhutan's Fourth King put Happiness in the collective consciousness in 1971, by declaring that improving Gross National Happiness (GNH) is a more suitable goal for the welfare of his people, and challenging the idea that increasing Gross Domestic Product (GDP) should be the ultimate goal. Many other cities and countries have since declared and followed similar goals. For example, the vision for Dubai to become the "Happiest City on Earth" was cemented through the launch of the *Happiness Agenda* by H.H Sheikh Mohammed Bin Rashid Al Maktoum, Vice-President & Prime Minister of the UAE, and Ruler of Dubai. This vision is undoubtedly noble, with many technical, social and psychological challenges (Al-Azzawi 2019). The UN had also subsequently commissioned the World Happiness Report, to systematically measure and rank countries according to their levels of well-being (Helliwell et al. 2015). Such re-focus has also recently become more pressing,

A. al-Azzawi (✉)
Smart Dubai, Dubai, UAE
e-mail: Ali.Alazzawi@smartdubai.ae

E. Azar and A. N. Haddad (eds.), *Artificial Intelligence in the Gulf*,
https://doi.org/10.1007/978-981-16-0771-4_9

due to the massive growth in populations of cities, where the global urban population is set to reach 75% of the world's population by 2050. Many cities and countries are therefore actively seeking happier cities. But can technology help?

The idea that technology can help people become happier is a new one. In particular, the combination of technology, good design, emotional design, and systematic academic study of user experience is relatively new (Al-Azzawi 2013; Norman 2003; Pavliscak 2019; Walter 2011). For example, recent research has resulted in a proposed framework to enable designers to follow in order to create positive experiences with technology (Calvo & Peters 2014). Further, the concept of *Digital Well-Being* has also had much attention (Floridi 2015).

More specifically, can AI help make people happy? Though there are many debates around this question, that take into account the eventual evolution of AI (Bostrom 2014; Tegmark 2017), with variously utopic and dystopic views, rather than look into the distant future, this chapter is about now and tries to take a literal approach to answer this question in terms of using AI in its current state towards increasing happiness and the factors that influence it. Looking at the wellness aspect of happiness rather than well-being in general, some researchers have attempted to chart the "Emerging Artificial Intelligence Wellness Landscape", pointing out opportunities and areas of ethical debate (Kostopoulos 2018). Some of these issues will be discussed below.

However, a basic question may be raised about the utility of AI, in the sense of how does it actually help drive better well-being? It is possible to demystify such a question if first the drivers of happiness are clarified, and in what contexts such utility is being examined, then map possible AI technologies to these contexts and drivers of happiness.

This chapter attempts to answer this question by outlining a pragmatic framework to help reach such visions towards happier cities, aiming to find the utility, as well as highlight some challenges of AI technologies towards happier lives from various perspectives; people's happiness needs, user experience context and themes within a happy city.

2 Happiness & Well-Being

The interest of ancient philosophers in the topic of well-being (also referred to as happiness) has been the subject of extensive academic study and therefore has an extensive body of theory, empirically-based knowledge and practical methods, as well as the focus of the United Nations' World Happiness Report (Helliwell et al. 2015). The report aimed at assessing and ranking happiness across the world. Acknowledging the subjective nature of happiness, the rankings are based on a single question, that essentially asks a sample of a country's population, how happy they are, using the Cantril Ladder question (Cantril 1965), which aims to elicit a person's general evaluation of happiness for their life as a whole. However, for the purpose of this chapter, it is more useful to look at the various aspects of life that are linked to happiness. The ABCDE model of happiness needs (Fig. 1), developed by Smart Dubai (Al-Azzawi 2019), is helpful in this regard, because it focuses on the needs of people, when sufficed, lead to increases in happiness. As such, this model gives a practical context to how technology could help increase happiness. The following describes a summary of each of these needs.

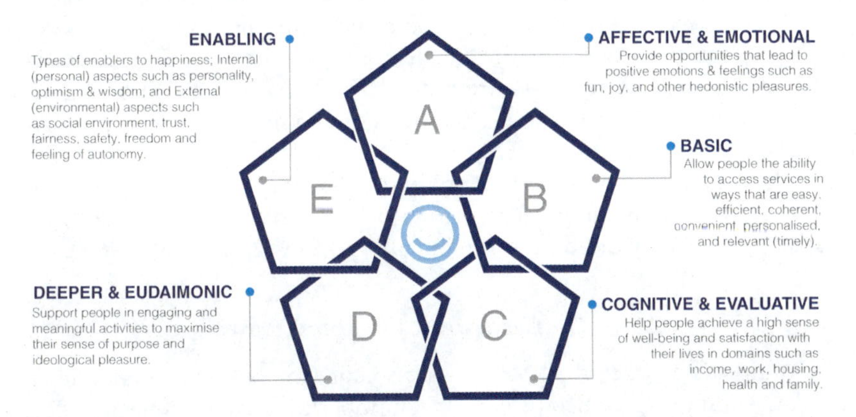

Fig. 1 The ABCDE model of Happiness Needs, as used in Dubai's Happiness Agenda (Al-Azzawi 2019). Reproduced with permission from Smart Dubai

2.1 A - *Affective Needs*

Diener and others have reported that "Subjective Well-Being is typically defined as the sum of affective and cognitive components" (Diener 1984; Miao et al. 2014; OECD 2013). Here, affect is meant as positive emotional states, that can be derived by various stimuli, such as simply having fun, perhaps enjoying a musical event. Benefits of positive affect can be directly related to happiness, where "evidence suggests that positive affect—the hallmark of well-being—may be the cause of many of the desirable characteristics, resources, and successes correlated with happiness" (Lyubomirsky et al. 2005). However, an important observation is that there is an optimum ratio of positive to negative emotions, where a "ratio of positive to negative affect at or above 2.9 will characterize individuals in flourishing mental health" (Fredrickson & Losada 2005).

2.2 B - *Basic Needs*

As highlighted by Maslow, people have needs that must be fulfilled in order for them to feel satisfied (1954). However, some of these needs can be translated into basic needs in terms of service provision in a city. People need basic access to services, in a usable and convenient way. Therefore, it is important to enhance the quality of experience people have with various services and technological interfaces (Al-Azzawi 2013; Stickdorn & Schneider 2011), and there are various design heuristics for such goals (Nielsen & Molich 1990). However, it is also important to note that people exhibit 'hedonic adaptation' as they adapt to various aspects of services, and eventually demand more in order to feel the same level of satisfaction as they did during the initial exposure (Kano et al. 1984).

2.3 C - *Cognitive & Evaluative Needs*

Cognition is a process related to assessment (thinking and evaluating), and according to Norman (2003) "cognition is a system for interpretation and understanding", which will eventually lead to an approach-avoidance response. With respect to well-being, this can be related, for example, to a person's assessment of their 'Satisfaction with Life' overall many life domains (Diener et al. 1985). A suitable list of such domains is one successfully used in life coaching methods (Biswas-Diener 2010, p. 104), which is based on an empirically-derived list of ten domains

(Frisch 2006). These domains are in the following general categories: professional, health, physical environment, relationships, and leisure. In the context of city life, it is important to add community, safety, and education. The full consolidated list of 12 life domains related to cognitive needs is therefore: *Income*, *Work*, *Commute*, *Learning & Education* (a person's sense of learning/skills opportunities and growth), *Friendships & Colleagues*, *Marriage/Partnership*, *Family*, *Community* (people and activities living around a person), *Housing*, *Safety*, *Physical & Mental Health*, and *Recreation*.

2.4 D - Deeper & Eudaimonic Needs

The value of having a sense of purpose and deeper meaning in life is a profoundly human trait (Seligman 2011; Zhang et al. 2016). Not only do people look for meaning in relationships, or actions, but also in stories and things (Csikszentmihalyi & Rochberg-Halton 1981). This sense has been highlighted and discussed by ancient philosophers like Aristotle. Referred to it as Eudaimonia (Huta 2014), this is about living a life of true virtue, where people are able to actualise their "daimon" (true self) (Vitterso 2003). Therefore, aside from functional needs, people also have deeper needs associated with deeper meaning, eudaimonia and purpose.

2.5 E - Enabling Needs

The needs outlined above were ones that a person could obtain or somehow able to arrange for themselves. In addition, there are two types of enabling needs; internal and external. These are ones that enable the above needs and enhance their positive value to a person. *Internal* (personal) enablers are factors that influence happiness, though tend to be about the person themselves or the way they react to their environment. Personality is no doubt an influencer (Steel et al. 2008), and the person's genetic predisposition is a significant factor (De Neve et al. 2012), as well as *personality traits* such as openness [to new experiences], conscientiousness, extraversion, agreeableness and neuroticism (McCrae & Costa 1987). However, there are also other personal factors that support and enable happiness such as optimism (Seligman 2006), and resilience (Seligman 2011), which are skills that can be learned (Lyubomirsky 2007). Also, a major enabler of happiness is having positive mental health (Dickerson 1993). The second type, *external* (environmental) enablers, is also important to happiness in terms of providing the right mix of environment. Aside from the benefits of the actual physical environment (including the natural environment), which may be considered part of the Cognitive & Evaluative needs, the social-environmental variables can also enable happiness, e.g. trust, governance, transparency, freedom, autonomy, fairness, as well as the general outlook for a culture, where some show general positivity and happiness (Hamilton et al. 2016). Other psychological dimensions relevant to the social environment have also been linked to well-being (e.g. autonomy, environmental mastery) (Ryff & Keyes 1995).

However, well-being may be explored at a wider scale. Aside from the life domains shown above (Cognitive & Evaluative Needs), there are also city domains or themes, that are linked to happiness. These were the subject of the *Happy Cities Agenda*, published by the Global Happiness Council, in their annual report on the Global Happiness & Well-Being Policy Report, highlighting worldwide best practice for improving happiness in society (Global Happiness Council 2018, 2019). The domains were split amongst two types: design-centric (e.g. urban design, nature, and quality of service) and enablers (e.g. trust, health, and sociality). At a different scale, the OECD developed the *Better Life Index* (BLI) to measure well-being at a country level, allowing progress in achieving it to be monitored. They propose two aspects, the first is *Quality of Life*, where they measure: health status, work-life balance, education and skills, social connections, civic engagement and governance, environmental quality, personal security and subjective well-being. The second aspect is *Material Conditions*, where they measure: income and wealth, jobs and earnings, and housing (OECD 2011). These structured themes provide a useful list of the kinds of focus areas where they may be useful to a city manager, wishing to explore the application of AI as a tool to increase happiness in the city.

3 Artificial Intelligence

Following Vannevar Bush's proposal of a system that amplifies people's knowledge and understanding (Bush 1945), Alan Turing made a serious proposal of how machines can simulate the intelligence of humans (Turing 1950). The seminal Dartmouth conference in 1956 ushered the AI era (where the term *Artificial Intelligence* was coined), with researchers and practitioners exploring the notion of AI and its possible benefits (McCarthy et al. 1955). The initial search was focused on modelling the way the brain worked, specifically with regard to general intelligence, and an accepted definition of AI, proposed by the co-founder of MIT's AI Lab in 1959, was proposed as "the science of making machines do things that would require intelligence if done by men" (Minsky 1968). Such definitions focused on general intelligence and encouraged exploration of how the mind works, rather than specific applications in society (Minsky 1985).

Today, we distinguish between two types of AI; Artificial General Intelligence (AGI), and Artificial Narrow Intelligence (ANI). AGI is concerned with reasoning and higher cognitive functions, while ANI, as the name suggests, is concerned with expertise and skills in a narrow field, like playing chess, or classifying and detecting patterns in images, text or speech. A more recent definition describes AI as "a system's ability to correctly interpret external data, to learn from such data, and to use those learnings to achieve specific goals and tasks through flexible adaptation" (Kaplan & Haenlein 2019). ANI uses techniques such as Machine Learning (ML), Deep Learning and Natural Language Processing (NLP) to develop models and gain insights from data. These methods can variously analyse static as well as dynamic data, in both, structured (e.g. tabular), and unstructured form (e.g. images (computer vision), audio, text). However, by far, most of the successes of AI have been in ANI and have become a common part of life, like the recommendations from Amazon, face recognition in cameras and Facebook, or sentiment analysis of social media. Any further reference to AI in this chapter is a reference to ANI, unless otherwise specified.

Still, AI has made significant progress, especially in the last decade. Aside from today's fictitious depictions and fantastic stories of emotional battles with embodied AI in movies such as *Ex Machina* (Garland 2014), and more seemingly prosaic tales from *Her*, telling the story of falling in love with a speaking operating system (Jonze 2013), early AI applications started with the logic solving program, *Logic Theorist* (1955), then a therapist called *Eliza* (1964), later IBM's *DeepBlue* beating a world chess champion (1997), moving on to IBM's *Watson* winning the verbal general knowledge game, *Jeopardy* (2011), and more recently DeepMind's *AlphaGo* (2016), beating the best human player in a notoriously difficult abstract strategy game. Significantly, AlphaGo was only given the game rules, and then it learned to strategise within the game and enhanced its performance, all by itself. However, for the majority of AI applications, at its essence, AI has been generally applied to skills that mimic and amplify distinct human abilities, such as perception & pattern recognition, classification & knowledge representation, reasoning & problem solving, learning (supervised & unsupervised), planning, and motion & manipulation (Russell & Norvig 2016). As a conceptual toolset, IBM's AI designers use their *AI Design Guidelines* (IBM 2019), to think about six core 'intents', or reasons, a person may have to use AI: accelerate research and discovery, enrich interactions, anticipate and pre-empt disruptions (monitor systems), recommendation (based on various parameters), scale expertise and learning (knowledge provision), and detect liabilities and mitigate risk (legal and formal document analysis).

Nonetheless, AI has been used towards particular types of outcomes, mostly through automation (as Andrew Ng puts it "AI is automation on steroids"). The following are some primary examples of outcomes, where AI has been applied:

- *Recommendation*: Based on comparing profiles of users (e.g. shoppers), the AI recommender modules are able to suggest choices based on the pattern of activity of other profiles with similar interests, thereby providing more personalised experiences of products and services. Also, related methods use pattern detection to generate new content (e.g. looking at the user's own content and behaviour in order to curate new content, as used by Facebook to create special postings on anniversaries or birthdays).
- *Interpretation & Understanding:* A key application of AI has been the interpretation and understanding of text, using NLP. The technology is used to detect and analyse various useful attributes, such as the language, topic, intent and sentiment.
- *Efficiency & Optimisation:* Since AI can be taught to map certain inputs to desired outputs, once placed in a feedback loop, it can be used to match the need and availability of resources. This kind of optimisation has been successful in the efficient delivery of goods and services. For example, Uber uses such tools to match the nearest driver to the waiting customer, taking into account distance and traffic. However, efficiency can also be used when allocating marketing expenditure towards specific segments of customers, or to optimise the design of a website, based on successful user-behaviour

criteria (e.g. A/B testing, leading to increased sales and enhanced customer experience).

- *Planning & Decision-Making:* AI has also been used to manage decisions based on previously successful outcomes. For example, deciding that a transaction is fraudulent based on a combination of variables, or providing a chatbot with the best response, based on how the chat is progressing. However, decision-making is also seen in motion planning for an autonomous vehicle, based on the perception of its environment, as well as tracking salient objects. In this case, the AI is also using models to predict the motion of other objects. Further, a common and highly beneficial application of AI has been in the health sector, augmenting decision-making in medical diagnosis.

These AI-based tasks have been used through a variety of channels (e.g. mobile apps, websites, speech), and in various manifestations; *direct*, where the user is expecting a specific function (e.g. an app that helps with travel/holiday options, or a pillow for better sleep), *indirect* (e.g. a shopping recommendation engine based on user behaviour, optimised mobility with a taxi, or health diagnosis at a hospital), or *embodied* in physical form (e.g. autonomous vacuum cleaner, or a manufacturing robot on an assembly line). This chapter describes practical ways where AI has been applied towards improving well-being, using the above manifestations, and various contexts of people's lives, though the details of AI technology and data analysis methods are beyond the scope of this chapter. However, there are challenges with deploying AI, touching on topics such as ethics and privacy, which influence well-being, and will, therefore, be briefly discussed in the 'Challenges' section of this chapter.

4 Applying AI to Happiness

This chapter follows a tradition in the AI field by exploring the application of AI towards well-being. Interestingly, one of the first applications of AI was a therapist called *Eliza*. The therapist allowed users to interact with a text-based interface towards counselling therapy (Weizenbaum 1966). Also, the combination of AI and well-being has received attention more recently, with some researchers concentrating on mental health. In one such article, the author examined the opportunities of further development, by focusing on the types of technologies used in various mental and physical health applications, exploring AI in terms of the examples of technical mediums of AI used towards well-being. These mediums were described as; *Intangible*: no physical form, running in the background and manifesting through notifications or signals, as required; *Tangible*: embodied by physical form (e.g. robot or vehicle), and; *Embedded*: fused into people's brain, through some type of Brain-Computer Interface (Kostopoulos 2018). Though it is useful to assess various AI applications in terms of technology and how it is used to help people, this view is perhaps more useful for anyone interested in benchmarking the technology landscape and how it applies to this domain.

However, as the extensive well-being literature has shown, there is much more to happiness and well-being than just mental and physical health. Nonetheless, it is also possible to explore different contexts of use of technology, where well-designed technology may support people's well-being. Some researchers have sought to show how the design of technology can be applied to various contexts (Peters et al. 2018), where they suggest the various 'spheres' of influence (e.g. behavioural, task, interactional). The proposed perspective builds on the Self-Determination Theory (SDT). This empirically-backed theory "identifies a small set of basic psychological needs deemed essential to people's self-motivation and psychological well-being" (Peters et al. 2018; Ryan et al. 2013). The three main psychological needs, *autonomy*, *competence* and *relatedness*, are mediated by *motivation*, *engagement* and *thriving/well-being*. In this case, ensuring that technology is designed for these mediators will help suffice these needs. The researchers then propose that these mediators can be enhanced for the different 'spheres': adoption, interface, tasks, behaviour, life and society. However, these spheres are essentially 'contexts' that can have wider scope about general consumption and use of technology, using the proposed model that is built "based on existing evidence for basic psychological need satisfaction, including evidence within the context of the workplace, computer games, and health" (Peters et al. 2018). Therefore, considering the Consumer Decision Process (CDP), and its simplified form (*Approach*, *Buying*, *Consummation* and *Divestment*) (Al-Azzawi 2013), 'approach' and initial 'awareness' of the technical product or service are also significant parts of the user experience, as this context is where early perceptions are born, and influence subsequent experiences. Also the city, as a whole ecosystem where the person is living, is the macro-context. Therefore, a more complete set of contexts includes: *awareness*, *adoption*, *interface*, *tasks*, *behaviour*, *life*, *society* and *city*.

However, with a more human-centric view (i.e., from the point of view anyone wishing to apply AI to a particular happiness challenge), which is ultimately what the goal is, it is possible to view AI tech from the perspective of needs for happiness (e.g. ABCDE needs of happiness, outlined earlier) (Al-Azzawi 2019). Alternatively, happiness from the perspective of living in the city may be explored, using themes outlined in previous studies, such as urban design, place-making, mobility, trust, economy, and sociality (Bin Bishr et al. 2019). In this way, an AI practitioner can choose the most appropriate perspective to achieve the happiness goals they have (Fig. 2).

Nonetheless, regardless of the perspectives, the quality of experience with technology is also important. This quality is mainly derived from seven major constructs that are salient to users and their experience of technology; novelty, usability, aesthetics, physicality, convenience, value and even complexity (Al-Azzawi 2013). However, though some technology can be inherently complex, good design can take complexity as a design constraint and is something to be taken into account, rather than ignored. Not to be confused with 'complicated' (which is a sense of being not understandable by the user (Norman 2011), complexity is everywhere, where there are many variables contributing to a system, and AI may be used to help in such cases. Still, at its simplest, AI methods are essentially modelling tools, that allow identification or prediction,

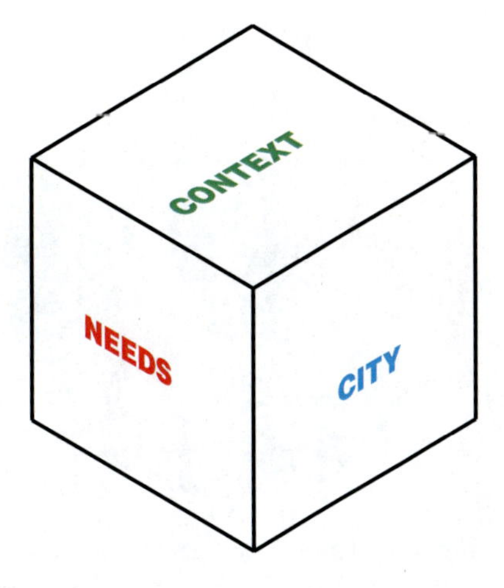

Fig. 2 Perspectives on experience with technology: People's Needs, User Context, Happy City

from an input A, to an output B. So, why use AI for happiness? What task/problem does AI try to address?

Considering using AI to enhance the quality of people's experiences from these perspectives provides the potential of improving well-being, efficiently and in ways that were not possible without AI. The following sections show many examples, across three perspectives, and they are cross-cutting types of manifestations in terms of mediums and methods.

4.1 Happy Cities Agenda

In the second of two consecutive publications by the Global Happiness Council, the Happy Cities Agenda (HCA) was presented in order to help city managers and custodians improve the quality of life for their citizens and visitors to their city (Bin Bishr et al. 2019). The HCA highlighted the main aspects of a city that have been empirically shown to be linked to well-being (based on consultation with experts and a literature review) and was outlined in the form of a practical tool, showing all the themes. The tool categorises the themes in two parts; *design* and *enablers* (Fig. 3). A full description of the data and rationale underpinning the tool is available in the document, at the Global Happiness Council's website (www.happinesscouncil.org). Therefore, for each of these themes, real and active examples are given in the remainder of this chapter, where AI has been

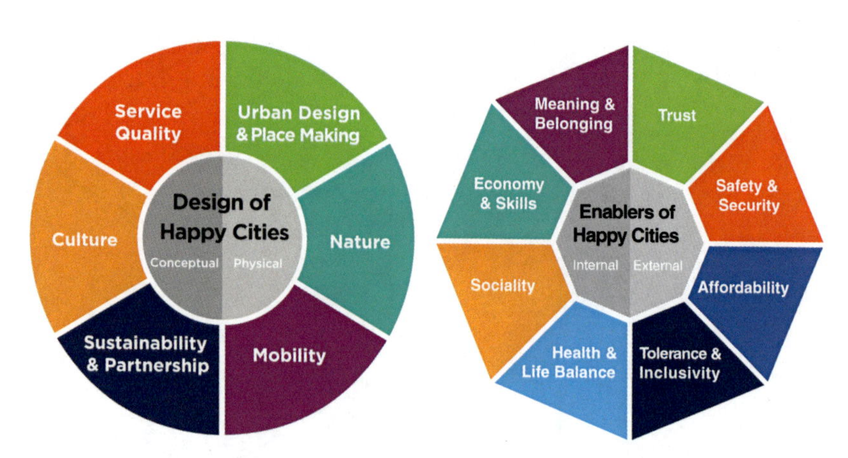

Fig. 3 Happy City Agenda tool published by the Global Council for Happiness and Well-being (2019). Reproduced with permission

usefully applied, leading to happier cities. In this context of AI, data is obviously a key element, and certainly, its use in cities is not new, where "computation is now routinely being embedded into the fabric and infrastructure of cities producing a deluge of contextual and actionable data which can be processed and acted upon in real-time" (Kitchin 2017). However, this extensive (by no means exhaustive) list of AI examples is not intended as a laundry list, but to give real working examples across the city, and to give a sense of the breadth of applications for happiness across the city. An important and consistent feature of these examples is the extensive use of data (big or otherwise). Nonetheless, although accessing data may be challenging in different ways, various methods of collaborations and mechanisms for sharing data are available.

5 Design of Happy Cities

5.1 *Urban Design & Place-Making*

The design of urban structures within the city has a profound influence on the well-being of its inhabitants (Gehl 2010). As such, attending to the quality of the design of various systems and infrastructures will be of value to anyone intending to work towards a happier city. There are many projects that use data and AI to get such value. How AI is and will influence cities is the first topic of exploration for a 100-year study on AI, based at Stanford (Stone 2016), looking at the influence of AI on North American cities. This focus is also evident in the private sector, where Alphabet's Sidewalks Labs use data and AI extensively to optimise the model development in Toronto, in order to ensure a high degree of livability in their urban development (Kwun 2018). Still, livability can also be measured retrospectively. In London, the StreetScore project systematically analyses street images and, based on many factors, assigns a score for liveability (StreetScore 2018). Similarly, the CHAOS platform also provides a *Liveability Index* in Finland (Laitala 2018). However, in Ho Chi Minh City (Vietnam), city managers have looked at the problem from space. They use free/low-cost satellite images to monitor the growth of urban zones in order to ensure sufficient resource planning data and manage growth (Goldblatt et al. 2018). These data, and others, have typically been used in city *Digital Twins*, in order to run simulations, aimed at improving the overall city design (Marr 2019).

5.2 *Nature*

There are many studies showing the well-being benefits of natural (green and blue) spaces, and many municipalities take an active approach to monitoring and increasing the amount of green space. For example, Dubai Municipality runs a collaborative project with the Mohammed Bin Rashid Space Centre to use high-quality satellite images to continuously measure the amount of green space in the city of Dubai, in order to manage their greening policies in the city. However, the topic of nature also includes ensuring the quality of the environment is free of pollutants. The agricultural machinery manufacturer, John Deere, acquired an AI company (Blue Valley) in order to improve the efficiency of pesticide and fertiliser use and protect the environment from the chemicals running into groundwater. The AI technology uses computer vision to precisely distinguish between weeds and crops, and ensures that only weeds are sprayed. This has led to a 90% reduction in fertiliser use, as well as an improved yield for the amount of land required for the target crop amount (Simonite 2017).

5.3 *Mobility*

Mobility is a key factor of living in the city and can be a source of frustration. Aside from the negative impact on economic costs, constant traffic delays can lead to a reduction in happiness (Stutzer & Frey 2008). It is therefore not surprising some taxi companies invest heavily in AI (e.g. Uber, to the extent that it is an 'AI first company', meaning that its business model is entirely based on AI). Uber uses data to optimise for pick-up times and locations, using road congestion data to give users an optimal experience (Reese 2016). However, other companies, such as Volvo and Tesla, use data to improve safety and user experience for private transport. Tesla uses data collected in the car, around the car (other fixed sensors), as well as data from wider scope (such as weather). The company uses these data to improve overall safety and has reported a drop in airbag deployment from 1.3 to 0.8 deployments/Million miles driven, due to the use of its Autopilot technology (Marr & Ward 2019; Marshall 2018), while Volvo has augmented its low-pollution electric/hybrid fleet with AI that helps monitor service requirements in order to predict spares inventory, and reduce waiting times for its customers (Morss 2018; Volvo 2017). However, public transport has also benefited from AI tools. In Seoul, the city managers used telco data, from 3Bn mobile calls, to analyse night bus routes. They found that the buses were not stopping at the most convenient and efficient places, for the people who actually used the buses. They consequently re-routed the buses, saving overall energy spend in the city, unnecessary taxi rides, increasing adoption of public transport and improving the quality of life for its citizens.

5.4 *Sustainability & Partnership*

One of the important aspects of raising the quality of life in a city is to do so in a sustainable way, and partnerships are a key method. In this way, all stakeholders are incentivised, with improved well-being as the outcome. In Latin America, Telefónica has demonstrated such partnership, along with the use of AI, in its project aimed at improving the quality of life for 100 M people. The company has partnered with Facebook, local authorities, local organisers and entrepreneurs to make the project happen. The project uses satellite images to identify inhabited areas as well as map out the transport infrastructure and existing communication towers, in order to deliver internet connectivity in the most efficient way. The project aims to deliver economic, educational as well as health benefits to a significant population, that would otherwise be left behind (Lopez 2018). However, such benefits can also be gained by smaller-scale projects where data and analysis are shared by symbiotic relationships such as food banks and restaurant chains, and employers and social organisations aiming to reduce unemployment. However, it is important to note that partnership is cross-cutting across all themes in the city, and not exclusive to domains outlined above, pooling resources/data towards value for the citizen.

5.5 *Culture*

Culture is part of the fabric of a city and has many benefits towards well-being, like a sense of shared meaning and belonging, and positive aesthetic value. Therefore, promoting cultural activities, including performance arts, is important. This can, of course, be done locally, but also using wider scope tools and products. Spotify is a well-known prolific platform providing a vast array of music for all tastes. The company has used AI to improve the listening experience of its users, as well as promote new artists. The approach has used typical methods of profiling users' listening habits and cross-promoting to other users with similar profiles. However, Netflix, a major provider of video content, has gone a step further. Though they also use profiling to provide content cross-promotion, they have also used AI to figure out a content formula. They used AI to identify features of successful content and have used this formula to create new content which will have a high chance of success. However, though such approaches have improved the overall customer experience, they have sought to improve the viewing experience by reducing the required bandwidth for streaming content. Using their proprietary AI method, they were able to reduce bandwidth requirements by 1000 times, leading to less churn (customers cancelling their subscription), and improved customer experience and brand value (Vena 2018).

5.6 *Quality of Service*

It is not surprising to note that the quality of service (e.g. speed, usability, and convenience, make people happier using a service). Indeed, customer satisfaction (CSAT) is a common proxy for the quality of service. For these reasons, many companies make quality a priority. For example, Verizon (a major telco in the USA) has sought to use AI to make improvements to their service (Knowledge@Wharton 2017; Marr 2018). The company has used its CSAT metrics, as well as 3 Gb/s of real-time data from its service infrastructure, to monitor and model its service provision. In this way, they are able to be proactive in dealing with potential failures. Interestingly, the AI tools allowed them to realise that their products were over-performing and allowed them to re-brand, for example, a 750 Mb product was actually providing 1 Gb of data throughput. Re-branding this service gave them more customers and increased their brand value in the marketplace. This kind of AI analysis was also used by Viacom (a major video content provider). They realised that the time it takes for the customer to see the first frame of the video, as well as the amount of buffering, were critical to keeping the customer engaged and not jumping off to a competitor (Databricks 2018). Knowing these to be their key quality features, they set about ensuring their business metrics were linked to these features. They were also able to see correlates with social media posts by their users and social influencers. Using these data, they ensured that their systems were able to dynamically re-allocate resources where required to maintain quality of their digital services. However, AI can be used in the real world not just the virtual world. Walmart has demonstrated an innovative and unusual application of AI to improve its services and keep customers happy. They used roaming robots with computer vision AI to ensure that their data was almost real-time regarding what items were on which shelf in their stores. This information was then made available to their customers in their mobile app, where they could navigate their large stores, being able to quickly and accurately locate items they were looking for. To date, Walmart has a small army of robots patrolling 50 of their stores, and giving valuable data to their supply chain and its customers, ensuring both business and customer value (Banker 2019).

6 Enablers of Happy Cities

6.1 *Trust*

Trust has been shown to be a correlate of well-being (Helliwell et al. 2016), and transparency and engagement can improve trust in government institutions. Trust may be defined as "a person's belief that another person or institution will act consistently with their expectations of positive behaviour" (OECD 2017). Therefore, any technology that increases trust in a society will lead to positive well-being. Social media companies, like Twitter and Facebook, have been working towards increasing trust in their platforms by using AI to combat fake news. However, one organisation that has sought to improve transparency at a local level is the Press Association (PA). They used AI to automatically generate high-quality local news because today's social media has made it commercially challenging to have local reporters assigned to small areas. Their AI trawls through all the news feeds they have, looking for news that is relevant to local regions. The AI then generates articles to be published locally, ensuring residents of small towns are kept abreast of local affairs, and feel connected and engaged with their town, thereby enhancing their sense of democracy (PA 2018). However, trust can also be enhanced in the commercial world by reducing the prevalence of counterfeit products in the market. A company called Entrupy produced a system that allows manufacturers, sellers and buyers to use hand-held devices (also possible to use add-on units to mobile phones), to scan objects with the camera (Sharma et al. 2017). The system then checks against its database of bona fide products through its network with suppliers. The company has reported a 98.5% accuracy in identifying genuine products, thereby making it difficult for the $1T counterfeit industry to continue with its trade. This system, therefore, gives sellers and buyers added confidence and trust in making commercial exchanges.

6.2 *Safety & Security*

There is no doubt that feelings of safety and security improve well-being, and these can manifest in many ways. In a traditional sense of safety in the public realm, many countries have employed AI to monitor places for suspicious activity and even face recognition to look for criminals. Amongst many, the Chinese company, SenseTime, has successfully applied such technologies to automatically raise an alarm if suspicious crowd movement is detected, as well as monitoring the movements of single individuals who may be loitering. Also, Dubai Police and Smart Dubai have announced their collaboration in monitoring crowd flow in the city (e.g. during festive seasons), so that the police can deploy resources in the most efficient way and maintain safety and security in the city. Still, safety and security can be at a closer scale, and companies like Apple have recognised this. For such concerns, they have made deliberate effort to protect the privacy of their smartphone users, and have, for example, ensured that fingerprint data is maintained at a hardware level and never leaves the phone. However, some personal and financial data must be shared, by the necessity of current financial systems, either on cards or through networks. American Express, amongst others, have used AI to protects its customers against fraud by looking for uncharacteristic patterns in card use (across a much wider pool of data), and generating a 'card-decline' signal to the merchant (Kumar 2015). However, falsely declining cards creates inconveniences for customers giving them unsatisfactory and bad experiences. Therefore, credit card companies have enhanced their AI to include more complex data gathering to include in their AI prediction model.

6.3 *Affordability*

Though cities are economic magnets for many, they can also become unaffordable to many. Therefore, any effort to ease financial burdens will have a positive influence on well-being. One of the applications of AI, in this regards, is in the smart meter technologies. For example, in 2018, Dubai's electricity and water supplier has smart meter coverage for 80% of its customer base (DEWA 2018) and uses the technology to help customers lower their costs of consumption, with features like notifications in case of unusual spending patterns. However, it is not just regular expenditures that can benefit from AI. The credit check company, Experian, re-uses its data to optimise the process of mortgage application (Marr & Ward 2019). This has significantly reduced the time, and exchange of redundant data, thereby reducing the cost of application for businesses, as well as the challenges to customers, caused by a notoriously difficult and stressful process.

6.4 *Tolerance & Inclusivity*

Social & economic inclusion and tolerance of others have positive well-being and economic value, in addition to being ethical. For these reasons, based at the University of Southern California and in collaboration with Los Angeles Homeless Services Authority (LAHSA), an AI-based project aims to improve the efficacy of providing homes to homeless people in Los Angeles (CAIS 2016). AI is used to address the challenge of housing the estimated 52,765 homeless people (in 2018) in LA (Harris Green 2019), and it does this by looking at the various data attributes of the people, as well as past cases of housing, and find the best fair and efficient options that have the least likelihood of a person returning to the streets. However, AI can also help social challenges in a more embodied manner, as has been demonstrated by Trinity College Dublin. They created a mobile robot, now in its second version as Stevie II, which is able to move around the home of an elderly person, interacting with them verbally, while also being able to control devices in the home, as well as the ability to call for help using the standard communications network (Roberts 2019). Another feature is that the sensors and the on-board camera are able to facilitate video chats or CCTV-like functions. This robot also provides some level of 'social' interaction to alleviate loneliness.

6.5 *Health & Life Balance*

A primary driver of happiness is health (both physical and mental), and there are many applications of AI in this field. AI has been able to help with detection of symptoms that are visible in medical images, such as X-Rays and MRIs, as well as being used in tracking patterns of vital signs, in order to alert medical practitioners and assist patients before a problem progresses, as being applied by the Dubai Health Authority. For example, the Chinese company, Infervision, has employed AI to make good quality diagnosis available to rural communities that have a lack of resources or expensive MRI equipment, and so rely on cheaper X-Ray and CT scanners (Shieber 2017). The AI tool they developed has helped detect early stages of lung cancer, to the extent that it is treatable, and are now working on developing AI tools towards helping stroke patients. Another example is illustrated by an ambitious project that was undertaken by Elsevier (academic publisher), after they recognised they were holding a significant proportion of the world's medical journals, over 140 years worth of publications. They set out to use AI to analyse the text, picking out treatments that worked, and combined these data with anonymised patient history data, as well as data from 5 million insurance claims, looking for successful outcomes. From this combination, they built their Advanced Clinical Decision Support platform (Marr & Ward 2019). The platform helps medical practitioners by suggesting an optimal decision, as a recommendation, and they will then have the options to follow. The approach has been successful by showing 85% adherence to the treatment pathways suggested by its oncology platform. However, AI can help with more prosaic, yet important aspects of health, like a wearable bracelet (e.g. Fitbit), to help analyse exercise, heart rate and sleep patterns, giving the user recommended lifestyle changes.

6.6 Sociality

Relatedness is a major and significant contributor to well-being. People need people, and ways that increase and improve the quality of relationships with others leads to improved well-being. Sociality is meant here as the "tendency of groups and persons to develop social links and live in communities…the quality or state of being social" (CollinsDictionary.com).

As discussed in the 'partnership' theme, Telefónica (Spanish telco), undertook to connect people in Latin America, who were living in areas that are difficult to reach. Such a project will undoubtedly lead to improved well-being directly, by allowing people to connect with each other easily, but also has other benefits, such as economic, educational and health. The company used AI to scan satellite images to find areas that are inhabited, yet had no connectivity. They further scanned the images to map the transport networks in order to find the lowest cost of supply and maintenance, as well as existing communication towers, in order to piggyback on existing structures, and further minimise project implementation costs. Their goal is to connect 100 M people who have been left behind and include them into the economy and the society at large. However, the social fabric of society can also be challenged in other ways. Many social media platforms have found a growing cyberbullying trend and have sought to address this problem. Instagram has used AI to analyse comments in their platform, scanning for negative content (text, video and audio), regarding race, gender and appearance (Instagram 2018). The AI uses advanced techniques to assess the semantics of the content, as well as being context-sensitive, where it can ignore banter and comments said in jest by friends.

6.7 Economy & Skills

A primary reason for people to move to a city is for economic opportunities, and these should be sought and enhanced, in order to increase and improve their quality and likelihood, for communities, businesses and individuals. There are many applications of AI to improve business performance, by profiling customers to improve marketing and sales, as Harley Davidson did with their motorcycle customers (Power 2017). However, Mastercard sought to reduce costs (cost of lost business to their merchants, and inconvenience to their customers). They used AI to profile a large set of data (e.g. user behaviour, product, seller, location), which created systems to reduce the chances of a card being falsely declined significantly. Still, another example of AI shows how it can be used to find new opportunities. Dubai-based service aggregation company, Mr Usta (www.MrUsta.com), uses its data to map demand for services across the city in the form of a heatmap (Fig. 4). For example, a laundry business or home repair company would use such a map to deploy its resources in the most efficient way, reducing callout times.

Jobs and skill-supply are of course other aspects of economic benefits in the city. In order to help with improving the process of matching people with relevant skills to employers who needed these skills, Linkedin has used AI in its social platform (Jersin 2017). They recognised the high cost of recruitment, and how hiring the wrong person can be a costly mistake. They, therefore, created AI to match potential-employees, recruiters and skills courses, and based on past data of successful matches; they are able to predict outcomes of people's application and suitability against

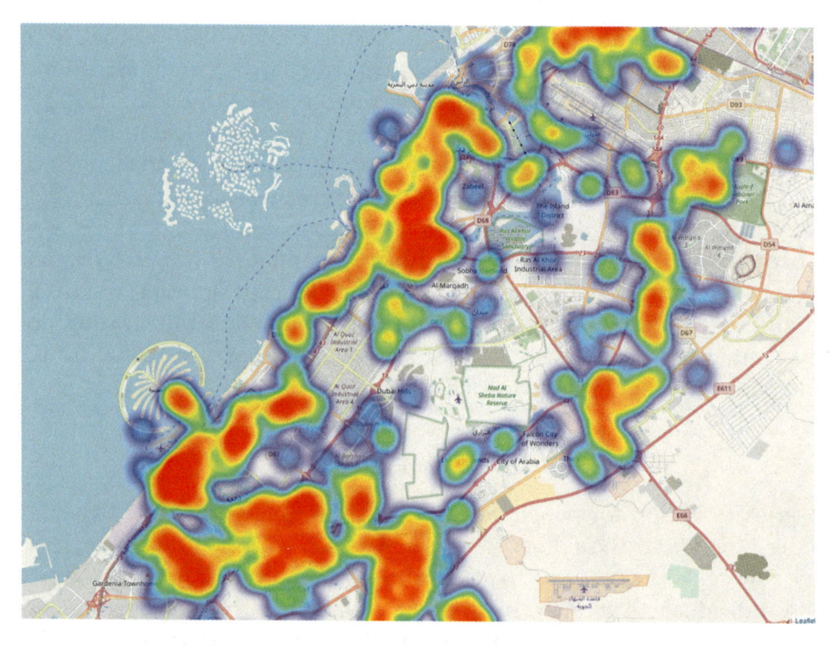

Fig. 4 Mr Usta heatmap overlay on the city of Dubai, showing areas of high demand for specific services. Reproduced with permission

job profiles. They did this using information that the recruiter could not easily see and analyse, like people's interests, and posts they liked. Their recommendation system was able to achieve an increase of 45% in person-to-person messages, as well as a twofold increase in the amount of engaged traffic between recruiters and employee. However, Unilever had an even more ambitious project. They created AI to read the CVs of applicants to their company and saved 70,000 h of manual processing for the initial screening (Feloni 2017). The same AI also gives automated assessments to the applicants, which was not economically possible before their system. They also created an AI-based chatbot to help new recruits find their way around HR rules, information and benefits, which got a customer satisfaction score of 3.9/5. However, all this automation can, of course, be a double-edged sword, and the OECD has published a paper discussing these issues, highlighting the jobs at most risk from AI-based automation (OECD 2018).

6.8 *Meaning & Belonging*

All social media platforms give users an opportunity to express themselves and enhance their sense of belonging by connecting with groups, friends and family. Facebook has become the iconic platform in this regard, especially due to its size and therefore making it more likely for people to find what they want. However, Facebook did not rely on standard marketing techniques to grow its user-base. The company used AI to analyse user behaviour and focused on its primary insight. They realised that people are most likely to continue to use their platform if a user connected with at least 7 people within 10 days of joining (Marr & Ward 2019). This became their primary KPI and target for all their recommendations and related on-platform marketing activities. However, AI is used much more extensively than user-behaviour analysis. They also use AI for face recognition of pictures that are uploaded, automatically tagging people which the user has already identified in previous posts, and subsequently use such identifications to find other people in the wider network, friends of friends, to increase the chances of success for the recommendations. Their face recognition AI has achieved 97.35% accuracy, higher than human skill. Still, Facebook analyses the text within its platform to model for anti-social behaviour, and even suicide prevention. Of course, there are also commercial benefits that Facebook capitalises on, with regard to better matching advertisers on its platform, based on people's interests and activities. But, with regard to making meaning, Disney is perhaps the most well-known for being the magic makers of the entertainment and leisure industry. They use AI to craft the most magical experience for their 'guests' at their massive parks. Disney creates a bracelet with an RFID tag, as part of their *MyMagic* + initiative, to track users as they move around the parks, giving them highly personalised recommendations and access to rides, that most fit their preferences (Kuang 2015). In this way, Disney maintains a friction-free experience with reduced queues, allowing its guests to focus on magical memories and experiences.

7 Needs & Contexts

As discussed earlier, there are different types of needs that, if sufficed, may lead to a person's happiness; Affective, Basic, Cognitive, Deeper, and Enablers. AI has been used across these different types of needs using different methods, with various degrees of sophistication. For example, in terms of affect and emotion, such as fun and entertainment, AI is used extensively in games in the entertainment and movie industries, for special effects, or modelling of game engines, and computer animation/computer-generated images (CGI). More recently, AI has also been used as a creative engine for art, like music and paintings. However, there are also some odd examples where AI is used to create positive emotions. In Japan, one particular hotel uses humanoids and dinosaur robots, as the hotel check-in 'staff', as well as a mobile robot to take luggage up to the guest's hotel room (Associated Press 2015).

For people's basic needs, there are countless examples of AI being used to enhance convenience and ease of use of systems, like the home speaker voice agents (e.g. Alexa and GoogleNow), as well as the smart thermostats (e.g. Nest) that use behavioural and environmental data in order to provide more economical and comfortable environment in an enclosed space, like an office or home. More recently, city concierge chatbots have appeared (e.g. Dubai's IBM/Watson-based *Rashid*), allowing city residents and visitors easy access to city services and information. Also, though it has now become ubiquitous and prosaic, AI-based recommendation engines in e-commerce channels serve an important role in providing a more efficient and personalised shopping experience (like Amazon). However, what is less common is to see hotels use AI to enhance the guest experience, as seen in some hotel groups (e.g. Jumeirah and Marriot). For Marriot, they used data they collect about their guest; before they arrive, from previous visits, as well as while they are staying. They use these data to set environmental preferences, such as room temperature and other factors, in order to create the most convenient and positive experience (EventMB 2018).

However, as discussed in the above sections, there are also several life domains where people have a cognitive and evaluative sense, and AI has been successful in improving well-being in many of these domains. For example, in commuting and mobility, self-driving and autonomous vehicles use AI extensively. Also, as discussed earlier, social media has a strong influence on supporting relationships between family, friends and communities. Furthermore, in the workplace, AI is being used to improve well-being and productivity. According to a survey of 3,000 workers in the UK, employees are expecting AI to improve well-being and productivity. Nearly half expect AI to benefit them, though 55% still fear that AI may take their jobs (Muller-Heyndyk 2018). However, one of the interesting and impactful applications of AI, in terms of wellness, has been in physical and mental health and is certainly a growing market that even Google is aiming at (Hill 2018). For physical health, aside from the plethora of on-body monitoring devices, wearable devices, such as Fitbit, are used to monitor physical activity, sleep quality, baselining heart rate measures, as well as helping to predict diabetes (Brown 2018); there are also devices such as smart pillows and sleep sensors that analyse and help improve sleep quality (Forbes 2018; ResMed 2018).

The above examples were mostly concerned with influencing or improving happiness. However, as with most things, "if you can't measure it, you can't improve it". Consequently, the happiness literature is replete with instruments and scales aimed at various psychological constructs, like 'satisfaction with life' or 'positive/negative emotions'. However, these are traditional methods, and many researchers have explored ways of measuring happiness, by using various types of AI methods, trained on different kinds of data (e.g. behavioural, economic, geospatial or motion data). One such example that combines traditional surveys with AI is Mappiness. This company provides a mobile app to users, simply asks them about their state of mind, analyses these data and maps them across a city, thereby giving a location and temporal description of well-being across a geographical area. However, in 2016, Smart Dubai went a little further and used Machine Learning to develop the *Happiness Algorithm* (Al-Azzawi 2019). This method used a large set of different types of city and personal data, including telco, financial, leisure activities and medical data (all anonymised and aggregated), to create a model that allows the prediction of a happiness score (verified to within a 10% error). In this way, the algorithm may be used to 'calculate' the average happiness of a group of people (e.g. based at a location or in an organisation), without asking them, but by merely running the prediction model using available behavioural and revealed objective data. Smart Dubai continues to work on implementing this tool across the city.

Such work that correlates well-being with objective measures (physical and non-physical) is related to the concept of the *Digital Phenotype* (Jain et al. 2015). In essence, it is the study of the behavioural and visible manifestations of the well-being of people and has been used to measure mental health with states such as depression, mood and emotions, through the data collected by sensors in a smartphone, including text entry, the accelerometers, camera and microphone (Dagum 2018; Insel 2017). AI has also been applied to manage addiction (Addicaid.com), and even attempts at replacing therapists with chatbots that give patients a more anonymised experience to help with anxiety (Therachat.io). Still, some applications of AI have even created embodied experiences, as mentioned above with Stevie-II (for the elderly), and even a lovable fury seal that learns its name. There is no doubt; there are many applications and opportunities of AI benefiting society, specifically as far as happiness and well-being. But, what are the challenges?

8 Challenges

So, what is the flip side of the AI coin? With all the power of AI and the benefits it can (and does) bring, AI still has its challenges and weaknesses (Ford 2018). Both inherent weaknesses, and in terms of implications of how AI may get used in the real world, just like any technology. For example, there are many examples of AI being used to generate text content, and some of these algorithms have been used by various nefarious organisations to spread fake news and fake comments on social media. Some of these have raised concerns that these kinds of activities may be used to influence public discourse, undermine trust and destabilise social systems like democracy. However, researchers have also demonstrated sophisticated AI being used to generate new artwork and more recently been used to show how a video can be created of a person speaking about topics they never spoke about—all fake (Suwajanakorn et al. 2017). However, AI itself is also vulnerable to attacks. Some researchers have demonstrated that 'minor perturbations' in the files of input images (too small to be noticed by the human eye) can mislead an AI algorithm to totally misinterpret an image, for example, classify a cat as guacamole (Ilyas et al. 2018). Still, there are also techniques being developed to fool AI computer vision (e.g. rendering a street "STOP" sign unreadable by an autonomous vehicle, yet still readable for a human) (Eykholt et al. 2018). These attacks show the potential for an 'arms race' between the good and bad forces of AI, where fraud and terrorism may be arenas for battle.

Other battles are being fought, where people's attention is the big prize (Wu 2016). AI is used intensively in such battles and accounts for much of the successes. For example, 70% of the increase in traffic on YouTube (an average of 60 min session for each visit) is due to its recommendation engine (Solsman 2018). These AI tools are being used to nudge people to un-well-being, for commercial benefits, by using shock

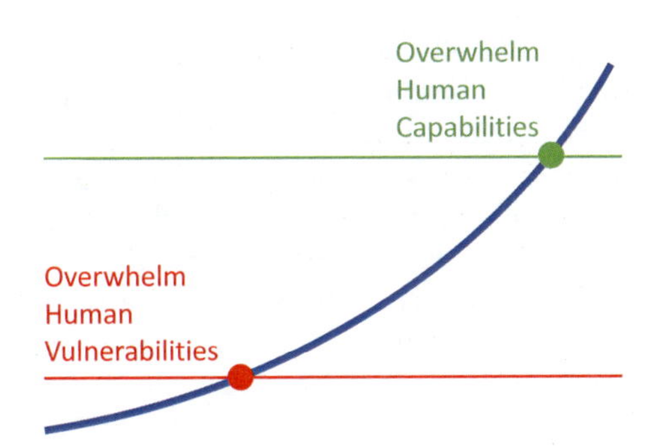

Fig. 5 As the capabilities of technology rise, the risk to humans from technology is at a lower level than usually discussed. Adapted from humanetech.com/problem (CHT 2019b)

and outrage as click-bait. The tools rely on exploiting people's weaknesses. This is important because "most recent conversations about the future focus on the point where technology surpasses human capability. But they overlook a much earlier point where technology exceeds human vulnerabilities" (CHT 2019b) (Fig. 5).

A further undesirable side effect of these attention-grabbing tactics is that they maintain people in a 'filter bubble' or 'echo chamber', where users are fed what is interesting to them, therefore a high likelihood of engagement, and more of what they wanted to see, and so act as reinforcement loops, rather than giving people other views and expanding their knowledge. In 2009, Google announced that it is providing "personalized search for everyone" (Google 2009; Sullivan 2009). This announcement marked a significant turning point, where major mass-content providers are providing "most relevant results" according to their algorithms, and their built-in inherent-biases, as well as people's own biases. However, people are mostly unaware of this phenomenon, where "in polls, a huge majority of us assume search engines are unbiased. But that may be just because they're increasingly biased to share our own views" (Pariser 2011), perhaps as a result of some kind of confirmation bias. Interestingly, this kind of loop can create "expectations of future [that] are haunted by the past behaviour" (Pavliscak 2019), which may trap people in a loop of their own making.

However, the AI algorithms with their relentless efforts to keep people engaged have also been instrumental in people's addiction and have been linked to an increase in depression and suicide due to increase in screen-time and social comparisons (Twenge et al. 2018). They have also manifested other psychological disorders, such as paranoia manifested as 'phantom vibrations', where users believe that the smartphone in their pocket has vibrated with an alert, but in fact, it did not (Kruger & Djerf 2017). One study showed cases of phantom vibrations in 89% of 290 undergraduates (Drouin et al. 2012), while other studies have singled out Facebook for 'iDisorders' (Rosen et al. 2013). Still, there are potential undesirable side-effects from the success of AI, that significantly help social media platforms grow. Some studies have shown that social media can isolate people or connect them; it depends on how people use it (Peters et al. 2018).

The important underlying concern here is that these potentially 'bad' algorithms are building blocks in an incremental and additive ecosystem. Highlighting the combinatorial evolutionary nature of technologies, Arthur points out that "Technologies inherit parts from the technologies that preceded them, so putting such parts together – combining them – must have a great deal to do with how technologies come into being" (Arthur 2009). Therefore, it is reasonable to assume that the DNA of our current technologies will persist, as future technologies inherit the current DNA. These algorithms are, therefore, the atoms of the bigger interconnected system that has no single designer. It is evolving based on the survival of the fittest, in a system that rewards and punishes certain traits. Many have outlined cogent arguments regarding the factors that must be taken into account as we build the *atoms* of tomorrow's "super intelligent" mind (Bostrom 2014), being careful of the inherent wisdom in handing over control explicitly, or implicitly, with its bad DNA intact.

A major potential Achilles heel in AI's DNA is the bias in its foundational data, which underpins its models. It is not a case of removing bias, but rather ensuring clarity about the bias and minimising unwanted or unfair bias. For example, biases in the sampling of the data, and the way the algorithms process the data have been known to accentuate cultural stereotypes. One study showed an unfortunate result when the AI was asked to show the parallel analogy between pairs of nouns, starting with "man is to computer programmer", the result was [it is like] "woman is to homemaker". Clearly, this is not actually true and highlights potential pitfalls resulting from bias in the programmer's choice of data and algorithm, as well as the sampling of the data itself (Bolukbasi et al. 2016). This is not an isolated example, as there are many datasets that have demonstrated such challenges, with recent privacy legal challenges against Mircrosoft, citing that "the most important datasets are rife with sexism and racism" (Gershgorn 2019), and more recently, the very method of collecting data for its Alexa product has led to Amazon being sued in two class-action lawsuits (Morgan 2019). These important issues have been highlighted by many, suggesting that such biases have created strong gender inequalities, treating "men as the default and women as atypical" (Criado Perez 2019).

For the above concerns and many more like privacy, as well as ensuring positive social value from AI, there has been a significant wave of discourse in setting deliberate strategies for AI, at various levels and contexts, including the ethics of design (IEEE 2017). For example, in the UAE, a Minister of AI was appointed at a federal level, to oversee the policies and strategies in the country. Also, Smart Dubai (the smart city organisation for Dubai) released guidelines on AI Principles (based on *Ethics*, *Security*, *Humanity*, *Inclusiveness*), as well as Ethics Guidelines, to ensure data and AI are used with attention to: *Fairness*, *Accountability*, *Transparency*, *Explainability* (Smart Dubai 2019). However, much of AI can be difficult to 'explain' as its mostly in the 'black boxes' of neural networks, though some researchers have pointed out the fallacy of trying to understand our complex and chaotic world, that has been revealed even more, by the utilitarian successes of AI over human efforts to model and conceptualise the various systems and sub-systems in our world (Weinberger 2019). Such views suggest the futility of trying to understand or explain the insides of the black boxes, rather than be sufficed with their utility.

Aside from the technical questions above, there are questions about values. How will we design machines to value what we value when we don't know ourselves? (Havens 2016). These issues discussed above also raise the question of paternalism. To what extent should the AI makers and regulators, decide what is good for people, and how much choice should they give people in deciding for themselves, without overwhelming them with choice, in other words, is it "legitimate for choice architects to try to influence people's behaviour in order to make their lives longer, healthier, and better."? Or is a softer approach more ethical, where *Libertarian Paternalism* "is a relatively weak, soft, and nonintrusive type of paternalism because choices are not blocked, fenced off, or significantly burdened" (Thaler & Sunstein 2009). Some advocates of 'humane technology' have outlined these problems (CHT 2019b) and proposed an agenda that strives to design technology to promote better well-being through better design ethics (CHT 2019a). There have also been efforts to render these ethical ideas into tangible technological systems and re-design our digital world with more human interfaces (Yuan 2019).

9 Conclusion

Many countries around the world have grasped the importance of AI, and some have set up specialised offices, as did the UAE's leadership by appointing the Minister of AI. These countries have sought to understand and capitalise on the rich potential and opportunities that this technology can bring, as well as addressing the challenges ahead. Much progress has therefore been made and drawing on a wide and eclectic sample of the various AI applications from Dubai, and other cities around the world, this chapter has presented the utility of AI towards making people happier in the city, from three perspectives. First from a view of the different themes in the city that influence happiness, then from a 'needs' point of view, when sufficed, will promote happiness, and finally from a point of view of the various contexts of use of technology.

However, there still remain some questions about what could be done to improve the focus of AI on well-being, rather than a tech or efficiency focus, or intended/perceived well-being. Many examples were given where the success of AI, from an engineering point of view, did not map to success from a humane point of view, and in fact had created negative effects (e.g. addiction or loneliness (Hari 2018). But, since the AI industry is busy creating ever more complex and capable pattern recognition modules, perhaps we are indeed on our way to creating a mind in the machine, one piece at a time. Nonetheless, regarding the application of AI in social settings, Eric Rice, the co-founder of *Center for Artificial Intelligence in Society* (CAIS), reminds us that "AI-driven work in the real world must happen at real-world speed… not at the pace of computer science" (Harris Green 2019). AI should be made good for the social context (i.e. made for society, with human values, and implemented at human pace). AI should be humanised. However, after considering the utility of AI, and if AI is indeed suitable as a solution to a happiness challenge, AI projects should aim for balanced and concurrent social, ecological and economic value, thereby contributing to a *socially smart city*.

Acknowledgements I would like to thank the many people who have enriched my understanding of AI, through various discussions and debates, especially my colleagues at Smart Dubai, and in the wider Government of Dubai. A special thanks goes to the leadership at Smart Dubai, for the unwavering support and vision that developed the Happiness Agenda into what it is today, continuously pushing and encouraging us to develop practical ways towards making Dubai a smart happy city. Most importantly, I feel a deep sense of gratitude to H.H. Sheikh Mohammed bin Rashid Al Maktoum, for the precious opportunity to contribute to the wonderful city of Dubai.

References

Al-Azzawi, A. (2013). Experience with Technology: Dynamics of User Experience of Mobile Media Devices. London: Springer.

Al-Azzawi, A. (2019). Dubai Happiness Agenda: Engineering the Happiest City on Earth. In W. A. Samad & E. Azar (Eds.), Smart Cities in the Gulf: Current State, Opportunities, and Challenges. London: Palgrave Macmillan.

Arthur, W. B. (2009). The Nature of Technology. London: Penguin.

Associated Press. (2015). Japan's robot hotel: a dinosaur at reception, a machine for room service. Retrieved from https://www.theguardian.com/world/2015/jul/16/japans-robot-hotel-a-dinosaur-at-reception-a-machine-for-room-service.

Banker, S. (2019). Walmart Expands Use of Bossa Nova's Robots from 50 to 350 Stores. Forbes. Retrieved from https://www.forbes.com/sites/stevebanker/2019/04/19/walmart-expands-use-of-bossa-novas-robots-from-50-to-350-stores/#4b3a9fa01f9b.

Bin Bishr, A., al-Azzawi, A., Baron, G., Yates, N., Montgomery, C., & Rodas, M. (2019). Happy Cities Agenda. In Global Happiness Policy Report. NY: Global Happiness Council.

Biswas-Diener, R. (2010). Practicing Positive Psychology Coaching: Assessment, Activities and Strategies for Success. New Jersey: John Wiley & Sons.

Bolukbasi, T., Chang, K.-W., Zou, J., Saligrama, V., & Kalai, A. (2016). Man is to Computer Programmer as Woman is to Homemaker? Debiasing Word Embeddings. Paper presented at the 30th Conference on Neural Information Processing Systems (NIPS 2016), Barcelona, Spain.

Bostrom, N. (2014). Superintelligence. Oxford: Oxford University Press.

Brown, A. (2018). Transforming diabetes through real-time data: Wearables + CGM. Retrieved from https://healthsolutions.fitbit.com/blog/transforming-diabetes-through-real-time-data-wearables-cgm/.

Bush, V. (1945). As We May Think. The Atlantic Monthly. Retrieved from https://www.theatlantic.com/magazine/archive/1945/07/as-we-may-think/303881/.
CAIS. (2016). Center for Artificial Intelligence in Society. Retrieved from https://www.cais.usc.edu/.
Calvo, R. A., & Peters, D. (2014). Positive Computing - Technology for Wellbeing and Human Potential. Cambridge MA: MIT.
Cantril, H. (1965). The pattern of human concerns. New Brunswick, NJ: Rutgers.
CHT. (2019a). Humane: A New Agenda for Tech. Center for Humane Technology, Retrieved from https://humanetech.com/newagenda/.
CHT. (2019b). The Problem. Center for Humane Technology, Retrieved from https://humanetech.com/problem/.
Criado Perez, C. (2019). Invisible Women: Data Bias in a World Designed for Men: Harry N. Abrams.
Csikszentmihalyi, M., & Rochberg-Halton, E. (1981). The meaning of things. Cambridge: Cambridge University Press.
Dagum, P. (2018). Digital biomarkers of cognitive function. npj Digital Medicine, 1(1). https://doi.org/10.1038/s41746-018-0018-4
Databricks. (2018). Customer Case Study: Viacom. AWS/Databricks. Retrieved from https://databricks.com/wp-content/uploads/2018/04/viacom-case-study.pdf.
De Neve, J.-E., Christakis, N. A., Fowler, J. H., & Frey, B. S. (2012). Genes, Economics, and Happiness. American Psychological Association, 5(4), 193–211. https://doi.org/10.1037/a0030292
DEWA. (2018). DEWA installs 595,755 smart water meters equivalent to 80.6% of Dubai's water meters. Retrieved from https://www.dewa.gov.ae/en/about-dewa/news-and-media/press-and-news/latest-news/2018/10/dewa-installs-595755-smart-water-meters-equivalent-to-806-of-dubais-water-meters.
Dickerson, A. E. (1993). The Relationship Between Affect and Cognition. Occupational Therapy in Mental Health, 12(1), 47–59.
Diener, E. (1984). Subjective Well-Being. Psychological Bulletin, 93, 542–575.
Diener, E., Emmons, R. A., Larson, R. J., & Griffin, S. (1985). The Satisfaction with Life Scale. Journal of Personality Assessment, 49, 71–75.
Drouin, M., Kaiser, D. H., & Miller, D. A. (2012). Phantom vibrations among undergraduates: Prevalence and associated psychological characteristics. Computers in Human Behavior, 28(4), 1490–1496. https://doi.org/10.1016/j.chb.2012.03.013.
EventMB. (2018). 3 Hotels Using Artificial Intelligence To Improve Guest Experience. Retrieved from https://www.eventmanagerblog.com/ai-concierge-hotel-guest-experience.

Eykholt, K., Evtimov, I., Fernandes, E., Li, B., Rahmati, A., Xiao, C., ... Song, D. (2018). Robust Physical-World Attacks on Deep Learning Models. Arxiv. Retrieved from https://arxiv.org/abs/1707.08945.

Feloni, R. (2017). Consumer-goods giant Unilever has been hiring employees using brain games and artificial intelligence - and it's a huge success. Business Insider. Retrieved from https://www.businessinsider.com/unilever-artificial-intelligence-hiring-process-2017-6.

Floridi, L. (Ed.) (2015). The Onlife Manifesto: Being Human in a Hyperconnected Era. London: Springer Open.

Forbes. (2018). This Smart Pillow Will Stop Your Partner's Terrible Snoring. Retrieved from https://www.forbes.com/sites/forbes-finds/2018/06/27/this-smart-pillow-will-stop-your-partners-terrible-snoring/.

Ford, M. (2018). Architects of Intelligence: The truth about AI from the people building it. Birmingham, UK: Packt.

Fredrickson, B. L., & Losada, M. F. (2005). Positive Affect and the Complex Dynamics of Human Flourishing. American Psychologist, 60(2), 678–686.

Frisch, M. B. (2006). Quality of life therapy: Applying a life satisfaction approach to positive psychology and cognitive therapy. Hoboken, NJ: John Wiley & Sons.

Garland, A. (2014). Ex Machina.

Gehl, J. (2010). Cities for people. London: Island Press.

Gershgorn, D. (2019). A Privacy Dustup at Microsoft Exposes Major Problems for A.I. Retrieved from https://onezero.medium.com/a-privacy-dustup-at-microsoft-exposes-major-problems-for-ai-53e0b4206e98.

Global Happiness Council. (2018). Happy Cities in a Smart World. In Global Happiness Policy Report. NY: Sustainable Development Solutions Network - UN.

Global Happiness Council. (2019). Happy Cities Agenda. In Global Happiness & Wellbeing Policy Report. NY: Sustainable Development Solutions Network - UN.

Goldblatt, R., Deininger, K., & Hanson, G. (2018). Utilizing publicly available satellite data for urban research: Mapping built-up land cover and land use in Ho Chi Minh City, Vietnam. Development Engineering, 3, 83–99.

Google. (2009). Personalized Search for everyone. Retrieved from https://googleblog.blogspot.com/2009/12/personalized-search-for-everyone.html.

Hamilton, K., Helliwell, J. F., & Woolcock, M. (2016). Social Capital, Trust and Well-being in the Evaluation of Wealth. NBER Working Paper, No. 22556. https://doi.org/10.3386/w22556.

Hari, J. (2018). Lost Connections: Why You're Depressed and How to Find Hope. London: Bloomsbury Publishing.

Harris Green, H. (2019). Could A.I. Help Get Homeless Youth Off the Streets? Retrieved from https://onezero.medium.com/could-a-i-help-get-homeless-youth-get-off-the-streets-f9fa9e53aeda.
Havens, J. C. (2016). Heartificial Intelligence: Embracing our humanity to maximise machines. NY: Penguin Random House.
Helliwell, J., Layard, R., & Sachs, J. (2015). World Happiness Report 2015: Sustainable Development Solutions Network - United Nations.
Helliwell, J. F., Huang, H., & Wang, S. (2016). New evidence on trust and well-being. NBER Working Paper, No. 22450. https://doi.org/10.3386/w22556.
Hill, L. (2018). Is Google Set To Disrupt The Market With AI Wellness Coach? Retrieved from https://www.welltodoglobal.com/is-google-set-to-disrupt-the-market-with-ai-wellness-coach/.
Huta, V. (2014). Eudaimonia. In S. A. David, I. Boniwell, & A. Conley Ayres (Eds.), The Oxford Handbook of Positive Psychology (pp. 201–213). Oxford: Oxford University Press.
IBM. (2019). Team Essentials for AI Workbook. Retrieved from https://www.ibm.com/design/thinking/page/toolkit.
IEEE. (2017). P7010 - Wellbeing Metrics Standard for Ethical Artificial Intelligence and Autonomous Systems. Retrieved from https://standards.ieee.org/project/7010.html.
Ilyas, A., Engstrom, L., Athalye, A., & Lin, J. (2018). Black-box Adversarial Attacks with Limited Queries and Information. AdrXiv.org, arXiv:1804.08598. Retrieved from https://arxiv.org/pdf/1804.08598.pdf.
Insel, T. R. (2017). Digital Phenotyping: Technology for a New Science of Behavior. Journal of American Medical Association, 318(13), 1215–1216. https://doi.org/10.1001/jama.2017.11295.
Instagram. (2018). Protecting Our Community from Bullying Comments. Instagram Press. Retrieved from https://instagram-press.com/blog/2018/05/01/protecting-our-community-from-bullying-comments-2/.
Jain, S. H., Powers, B. W., Hawkins, J. B., & Brownstein, J. S. (2015). The Digital Phenotype. Nature Biotechnology, 33, 462–463.
Jersin, J. (2017). How LinkedIn Uses Automation and AI to Power Recruiting Tools. Linkedin. Retrieved from https://business.linkedin.com/talent-solutions/blog/product-updates/2017/how-linkedin-uses-automation-and-ai-to-power-recruiting-tools.
Jonze, S. (2013). Her.
Kano, N., Nobuhiku, S., Fumio, T., & Shinichi, T. (1984). Attractive quality and must-be quality. Journal of the Japanese Society for Quality Control (in Japanese), 14(2), 39–48.
Kaplan, A., & Haenlein, M. (2019). Siri, Siri, in my hand: Who's the fairest in the land? On the interpretations, illustrations, and implications of artificial

intelligence. Business Horizons, 62(1), 15–25. https://doi.org/10.1016/j.bushor.2018.08.004.

Kitchin, R. (2017). Data-Driven Urbanism. In R. Kitchin, T. P. Lauriault, & G. McArdle (Eds.), Data and the City. Abingdon: Routledge.

Knowledge@Wharton. (2017). Tapping AI: The Future of Customer Experience at Verizon Fios. UPenn. Retrieved from https://knowledge.wharton.upenn.edu/article/competing-with-the-disruptors-a-view-of-future-customer-experience-at-verizon-fios/.

Kostopoulos, L. (2018). The Emerging Artificial Intelligence Wellness Landscape: Opportunities and Areas of Ethical Debate. Retrieved from https://medium.com/@lkcyber/the-emerging-artificial-intelligence-wellness-landscape-802caf9638de.

Kruger, D. J., & Djerf, J. M. (2017). Bad vibrations? Cell phone dependency predicts phantom communication experiences. Computers in Human Behavior, 70, 360–364. https://doi.org/10.1016/j.chb.2017.01.017.

Kuang, C. (2015). Disney's $1 Billion Bet on a Magical Wristband. Wired. Retrieved from https://www.wired.com/2015/03/disney-magicband/.

Kumar, N. (2015). New Age Fraud Analytics: Machine Learning on Hadoop. Mapr. Retrieved from https://mapr.com/blog/new-age-fraud-analytics-machine-learning-hadoop/.

Kwun, A. (2018). Our first look at Alphabet's smart city. Retrieved from https://www.fastcompany.com/90219315/our-first-look-at-alphabets-smart-city.

Laitala, A. (2018). AI and City Planning. Maankäyttö (Land-Use). Retrieved from http://www.maankaytto.fi/arkisto/mk418/mk418_extra_laitala.pdf.

Lopez, P. (2018). How Telefónica uses artificial intelligence and machine learning to connect the unconnected. Retrieved from https://www.telefonica.com/en/web/public-policy/blog/article/-/blogs/how-telefonica-uses-artificial-intelligence-and-machine-learning-to-connect-the-unconnected.

Lyubomirsky, S. (2007). The How of Happiness: A new approach to getting the life you want. London: Penguin.

Lyubomirsky, S., King, L., & Diener, E. (2005). The benefits of frequent positive affect: does happiness lead to success? Psychological Bulletin, 131(6), 803–855. https://doi.org/10.1037/0033-2909.131.6.803.

Marr, B. (2018). The Amazing Ways Verizon Uses AI And Machine Learning To Improve Performance. Forbes. Retrieved from https://www.forbes.com/sites/bernardmarr/2018/06/22/the-amazing-ways-verizon-uses-ai-and-machine-learning-to-improve-performance/#50056e37638a.

Marr, B. (2019). 7 Amazing Examples of Digital Twin Technology In Practice. Retrieved from https://www.forbes.com/sites/bernardmarr/2019/04/23/7-amazing-examples-of-digital-twin-technology-in-practice/.

Marr, B., & Ward, M. (2019). Artificial Intelligence in Practice - How 50 successful companies used AI and machine learning to solve problems. Chichester: Wiley.

Marshall, A. (2018). Tesla's Favorite Autopilot Safety Stat Just Doesn't Hold Up -. Wired. Retrieved from https://www.wired.com/story/tesla-autopilot-safety-statistics/.

Maslow, A. H. (1954). Motivation and Personality. New York: Harper.

McCarthy, J., Minsky, M. L., Rochester, N., & Shannon, C. E. (1955). A proposal for the Dartmouth Summer Research Project on Artificial Intelligence. Retrieved from http://www-formal.stanford.edu/jmc/history/dartmouth/dartmouth.html.

McCrae, R. R., & Costa, P. T. J. (1987). Validation of the five-factor model of personality across instruments and observers. Journal of Personality and Social Psychology, 52(1), 81–90.

Miao, F. F., Koo, M., & Oishi, S. (2014). Subjective Well-Being. In S. A. David, I. Boniwell, & A. Conley Ayres (Eds.), The Oxford Handbook of Positive Psychology (pp. 174-184). Oxford: Oxford University Press.

Minsky, M. L. (1985). The Society of Mind. NY: Simon & Schuster.

Minsky, M. L. (Ed.) (1968). Semantic Information Processing. Cambridge, MA: MIT Press.

Morgan, R. (2019). Amazon sued for recording children's voices via Alexa. Retrieved from https://nypost.com/2019/06/13/amazon-sued-for-recording-childrens-voices-via-alexa/.

Morss, J. (2018). The rubber hits the road with AI implementations. CIO. Retrieved from https://www.cio.com/article/3297496/the-rubber-hits-the-road-with-ai-implementations.html.

Muller-Heyndyk, R. (2018). Workers turning to AI to improve wellbeing and productivity. Retrieved from https://www.hrmagazine.co.uk/article-details/workers-turning-to-ai-to-improve-wellbeing-and-productivity.

Nielsen, J., & Molich, R. (1990). Heuristic evaluation of user interfaces. Proceedings of ACM CHI'90, 249–256.

Norman, D. A. (2003). Emotional design: Why we love (or hate) everyday things. New York: Basic Books.

Norman, D. A. (2011). Living with Complexity. Cambridge MA: MIT Press.

OECD. (2011). OECD Better Life index. Retrieved from http://www.oecdbetterlifeindex.org/.

OECD. (2013). OECD Guidelines on Measuring SubjectiveWell-being: OECD Publishing.

OECD. (2017). OECD Guidelines on Measuring Trust. OECD Publishing. https://doi.org/10.1787/9789264278219-en.

OECD. (2018). Putting faces to the jobs at risk of automation. Retrieved from https://www.oecd.org/employment/Automation-policy-brief-2018.pdf.

PA. (2018). More than 1,000 UK regional news titles now have access to stories jointly written by journalists and AI as RADAR launches new website. Press Association. Retrieved from https://pa.media/2018/06/18/more-than-1000-uk-regional-news-titles-now-have-access-to-stories-jointly-written-by-journalists-and-ai-as-radar-launches-new-website/.

Pariser, E. (2011). The Filter Bubble: What the Internet is hiding from you. London: Penguin.

Pavliscak, P. (2019). Emotionally Intelligent Design. CA: O'Reilly.

Peters, D., Calvo, R. A., & Ryan, R. M. (2018). Designing for Motivation, Engagement and Wellbeing in Digital Experience. Frontiers in Psychology, 9(797). https://doi.org/10.3389/fpsyg.2018.00797.

Power, B. (2017). How Harley-Davidson Used Artificial Intelligence to Increase New York Sales Leads by 2,930%. Harvard Business Review. Retrieved from https://hbr.org/2017/05/how-harley-davidson-used-predictive-analytics-to-increase-new-york-sales-leads-by-2930.

Reese, H. (2016). How data and machine learning are 'part of Uber's DNA'. Tech Republic. Retrieved from https://www.techrepublic.com/article/how-data-and-machine-learning-are-part-of-ubers-dna/.

ResMed. (2018). S + Sleep Sensor. Retrieved from https://www.resmed.com/us/en/consumer/s-plus.html.

Roberts, A. (2019). Meet Stevie II - Ireland's first AI robot designed to help care for older people. Retrieved from https://jrnl.ie/4636173.

Rosen, L. D., Whaling, K., Rab, S., Carrier, L. M., & Cheever, N. A. (2013). Is Facebook creating "iDisorders"? The link between clinical symptoms of psychiatric disorders and technology use, attitudes and anxiety. Computers in Human Behavior, 29(3), 1243–1254. https://doi.org/10.1016/j.chb.2012.11.012

Russell, S., & Norvig, P. (2016). Artificial Intelligence: A Modern Approach (Third Edition ed.). Harlow, England: Pearson.

Ryan, R. M., Curren, R. R., & Deci, E. L. (2013). What humans need: Flourishing in Aristotelian philosophy and self-determination theory. In A. S. Waterman (Ed.), The Best Within Us: Positive Psychology Perspectives on Eudaimonia. Washington DC: American Psychological Association.

Ryff, C. D., & Keyes, C. L. M. (1995). The structure of psychological well-being revisited. Journal of Personality and Social Psychology, 69(4), 719–727. https://doi.org/10.1037/0022-3514.69.4.719

Seligman, M. E. P. (2006). Learned Optimism: How to change your mind and your life. NY: Vintage.

Seligman, M. E. P. (2011). Flourish: A New Understanding of Happiness and Well-Being - and How to Achieve Them. London: Nicholas Brealey Publishing.

Sharma, A., Srinivasan, V., Kanchan, V., & Subramanian, L. (2017). The Fake vs Real Goods Problem: Microscopy and Machine Learning to the Rescue. Paper presented at the KDD '17, Halifax, NS, Canada.

Shieber, J. (2017). Chinese startup Infervision emerges from stealth with an AI tool for diagnosing lung cancer. Techcrunch. Retrieved from https://techcrunch.com/2017/05/08/chinese-startup-infervision-emerges-from-stealth-with-an-ai-tool-for-diagnosing-lung-cancer/.

Simonite, T. (2017). Why John Deere Just Spent $305 Million on a Lettuce-Farming Robot. Wired. Retrieved from https://www.wired.com/story/why-john-deere-just-spent-dollar305-million-on-a-lettuce-farming-robot/.

Smart Dubai. (2019). AI Principles and Ethics. Retrieved from https://www.smartdubai.ae/initiatives/ai-principles-ethics.

Solsman, J. E. (2018). YouTube's AI is the puppet master over most of what you watch. Retrieved from https://www.cnet.com/news/youtube-ces-2018-neal-mohan/.

Steel, P., Schmidt, J., & Shultz, J. (2008). Refining the Relationship Between Personality and Subjective Well-Being. Psychological Bulletin, 134(1), 138–161. https://doi.org/10.1037/0033-2909.134.1.138.

Stickdorn, M., & Schneider, J. (2011). This is Service Design Thinking. Amsterdam: BIS Publishers.

Stone, P. (2016). One Hundred Year Study on Artificial Intelligence (AI100). Retrieved from Stanford: https://ai100.stanford.edu/2016-report.

StreetScore. (2018). A tool to measure the quality of a place. Retrieved from http://dev.createstreets.com/wp-content/uploads/2018/07/StreetScore.pdf.

Stutzer, A., & Frey, B. S. (2008). Stress that Doesn't Pay: The Commuting Paradox. Scandinavian Journal of Economics, 110(2), 339–366. https://doi.org/10.1111/j.1467-9442.2008.00542.x.

Sullivan, D. (2009). Google Now Personalizes Everyone's Search Results. Retrieved from https://searchengineland.com/google-now-personalizes-everyones-search-results-31195.

Suwajanakorn, S., Seitz, S. M., & Kemelmacher-Shlizerman, I. (2017). Synthesizing Obama: Learning Lip Sync from Audio. ACM Transactions on Graphics, 36(4). Retrieved from http://grail.cs.washington.edu/projects/AudioToObama/.

Tegmark, M. (2017). Life 3.0 - Being human in the age of Artificial Intelligence.

Thaler, R. H., & Sunstein, C. R. (2009). Nudge: Improving Decisions About Health, Wealth and Happiness London: Penguin.

Turing, A. M. (1950). Computing Machinery and Intelligence. Mind, 49, 433–460.

Twenge, J. M., Joiner, T. E., Rogers, M. L., & Martin, G. N. (2018). Increases in Depressive Symptoms, Suicide-Related Outcomes, and Suicide

Rates Among U.S. Adolescents After 2010 and Links to Increased New Media Screen Time. Clinical Psychological Science, 6(1), 3–17. https://doi.org/10.1177/2167702617723376.

Vena, D. (2018). Netflix Streaming Gets an AI Upgrade. Motley Fool. Retrieved from https://www.fool.com/investing/2018/03/15/netflix-streaming-gets-an-ai-upgrade.aspx.

Vitterso, J. (2003). Flow versus life satisfaction: A projective use of cartoons to illustrate the difference between the evaluation approach and the intrinsic motivation approach to subjective quality of life. Journal of Happiness Studies, 4, 141–167.

Volvo. (2017). Volvo Cars and Autoliv team up with NVIDIA to develop advanced systems for self-driving cars. Volvo. Retrieved from https://www.media.volvocars.com/global/en-gb/media/pressreleases/209929/volvo-cars-and-autoliv-team-up-with-nvidia-to-develop-advanced-systems-for-self-driving-cars.

Walter, A. (2011). Designing for Emotion. New York: A Book Apart.

Weinberger, D. (2019). Everyday Chaos. Boston MA: Harvard Business Review Press.

Weizenbaum, J. (1966). ELIZA - a computer program for the study of natural language communication between man and machine. Communications of the ACM, 9(1), 36–45. https://doi.org/10.1145/365153.365168.

Wu, T. (2016). The Attention Merchants: The Epic Scramble to Get Inside Our Heads. New York: Knopf.

Yuan, J. (2019). Introducing Mercury OS. Retrieved from https://uxdesign.cc/introducing-mercury-os-f4de45a04289.

Zhang, H., Sang, Z., Chen, C., Zhu, J., & Deng, W. (2016). Need for Meaning, Meaning Confusion, Meaning Anxiety, and Meaning Avoidance: Additional Dimensions of Meaning in Life. Journal of Happiness Studies, 1–22.

CHAPTER 10

Women and the Fourth Industrial Revolution: An Examination of the UAE's National AI Strategy

Victoria Heath

1 Introduction

Although industrial progress may be relegated to just a few lectures in the class of world history, the speed at which that progress was made stretched across centuries. What has been dubbed the "Fourth Industrial Revolution (4IR)" however is expected to happen in just a manner of decades. In fact, it has already begun and the COVID-19 pandemic has only quickened its pace. The 4IR is changing every aspect of modern society by "blurring the lines between the physical, digital, and biological spheres" thanks to the exponential advancement and adoption of technologies, such as artificial intelligence (AI) (Schwab 2016). As Klaus Schwab, Founder and CEO of the World Economic Forum and author of *The Fourth Industrial Revolution*, wrote in 2016:

V. Heath (✉)
Montreal AI Ethics Institute, Toronto, ON, Canada
e-mail: victoria.heath@mail.utoronto.ca

E. Azar and A. N. Haddad (eds.), *Artificial Intelligence in the Gulf*,
https://doi.org/10.1007/978-981-16-0771-4_10

> The speed of current breakthroughs has no historical precedent. When compared with previous industrial revolutions, the Fourth is evolving at an exponential rather than a linear pace. Moreover, it is disrupting almost every industry in every country. And the breadth and depth of these changes herald the transformation of entire systems of production, management, and governance (Ibid.).

The 4IR is an extension of the First, Second, and Third Industrial Revolutions. The First stretched from the late eighteenth century until the beginning of the nineteenth century and saw the rise of mechanisation. The Second began at the end of the nineteenth century until the early twentieth century, characterised by the widespread adoption of new energy sources such as gas, oil, and electricity, as well as the development of telecommunication systems. The Third began with the rise of computers in the late 1960s and eventually saw the mass adoption of digital technologies and the Internet of Things (IoT) (Daemmrich 2017). The Fourth (or the 4IR) builds particularly on the Third Revolution's "data-centric foundations" (Davis and Philbeck 2019).

The impact of the 4IR on global competition and the international balance of power depends on how organisations and governments decide to utilise the technologies associated with the 4IR, particularly AI. Due to the fact that AI is a general-purpose technology, it has the potential to touch on almost every aspect of modern society, from making our smartphones even smarter to altering the nature of warfare due to the proliferation of lethal autonomous weapons. Who will dominate in this new era? Paul Scharre, senior fellow with the Center for a New American Security (CNAS) and Michael Horowitz, professor at the University of Pennsylvania, argue that the dominant nations will be those "with access to the best data, computing resources, human capital, and process of innovation…"(Scharre and Horowitz 2018). Further, a 2018 CNAS report detailed the key elements of national power in the 4IR, these include:

- Owning large quantities of the right type of data
- Training, sustaining, and enabling an AI-capable talent pool
- Computing resources
- Organisation incentivised and aligned to adopt AI effectively
- Public-private cooperation
- The willingness to act (Horowitz et al. 2018)

Another key element, implicit in those outlined above, is the inclusion of women in the development, deployment, and governance of AI. Not only is the inclusion of women a matter of national economic and military dominance in the 4IR, but more importantly, it is a matter of ethics and humanity. We risk creating AI that will exacerbate existing social injustices, such as racism, wealth inequality, religious discrimination, and more, if we don't ensure a range of perspectives, lived experiences, and expertise are included at all steps of the development and deployment process—starting with all genders. As stated in the WEF's 2018 *Global Gender Gap* report:

> The age of the Fourth Industrial Revolution (4IR) brings about unprecedented opportunities as well as new challenges. To take full advantage of new technologies, we need to place emphasis on what makes us human: the capacity to learn new skills as well as our creativity, empathy and ingenuity. By developing our unique traits and talents, humanity can cope with increasingly fast technological change and ensure broad-based progress for all. The equal contribution of women and men in this process of deep economic and societal transformation is critical. (World Economic Forum 2018b)

Governments across the world are jockeying for power in this new era by developing national AI strategies. To date, however, only 28 countries and/or regions have *coordinated* national AI strategies, and despite the importance of "owning the right type of data" and "enabling an AI-capable talent pool," which undoubtedly must include women, no single national AI strategy outlines women's inclusion, or gender more broadly, as an exclusive strategic priority (Kung et al. 2020). Unfortunately, adequate data and theories regarding this topic are not yet available due to the fact that most national AI strategies have only been made public since 2017–2018 and are vague at best. Thus, academic research into the political and national strategies around AI beyond national security considerations is scant. It's important to note, however, that several research centres such as the Future of Humanity Institute at the University of Oxford (Dafoe 2018) and the Canadian Institute for Advanced Research (CIFAR), are beginning to examine national AI strategies more thoroughly.

> Is ensuring the participation of women in the development, deployment, and governance of artificial intelligence (AI) vital to the future success,

security, and prominence of nations in the 4IR? If so, should women's inclusion in AI be considered an exclusive priority in national AI strategies?

I conclude that women's inclusion in AI is a significant factor in ensuring the successful development, deployment, and governance of AI, and therefore should be included in national AI strategies. This is for a variety of reasons which can be grouped into three broad categories, outlined below:

- Social—AI has the potential to exacerbate and entrench existing bias and inequality, particularly gender and racial inequality
- Security—Due to the fact that AI is a general-purpose technology, it has the potential to impact every aspect of society, and this can have a major impact on national and international security
- Economic—As governments jockey for power in the 4IR, they need a workforce with the skills and abilities to match labour market demands, this includes AI skills and expertise

For this research, I focus particularly on the economic category and the importance of including women in strategies to "train, sustain, and enable an AI-capable talent pool," one of the key elements of national power in the 4IR listed by Scharre and Horowtiz (Horowitz et al. 2018). I examine the research questions from a qualitative approach utilising academic research from a variety of fields, such as security, business, science, technology, women and gender studies as well as relevant news articles and data from government and intergovernmental organisations, to understand the current status of women in AI and why women's inclusion is important.

In the first section of this chapter, I will examine more broadly the status of women in the 4IR and why women's inclusion is important. Then, I will briefly examine the current state of national AI strategies and their priorities as categorised by CIFAR, and finally, proceed to examine the United Arab Emirates (UAE) as a specific case study in order to further contextualise this research (Dutton et al. 2018). The use of the UAE as a case study, or the use of a case study at all, is important due to the fact that to-date, national AI strategies are quite varied and difficult to study from a broad perspective. Therefore, examining a specific case study offers a better analysis of the research questions. In particular, the UAE is a useful case study because it has not only made AI a significant national strategic priority—evident in the creation of its *UAE Strategy for Artificial Intelligence* and the world's first Minister of State for Artificial Intelligence in 2017—but it has also indicated that the "empowerment of women" is a national strategic priority with the launch of the *National Strategy for the Empowerment of Emirati Women* in 2015. The UAE has also led the way in the Gulf region in the pursuit of both of these priorities by establishing advisory councils, government initiatives, and allocated funding for programmes.

In my conclusion, I suggest that a potential strategy to include women in national AI strategies is "gender mainstreaming," defined as "the deliberate consideration of gender in all stages of program and policy planning, implementation and evaluation, with a view to incorporate the impact of gender at all levels of decision-making" (UNESCO 2017).

"Women's inclusion" in this chapter refers specifically to the active integration of individuals who identify as women in the development, deployment, and governance of AI. "Development" refers to the AI talent pool that is developing the technical aspects of this technology (such as facial recognition, predictive analytics, and speech processing) and is the main focus of the UAE case study. "Deployment" refers to the utilisation of AI technologies by governments, organisations, companies, civil society groups, etc., in their operations, strategies, products, services, and/or other activities. "Governance" refers to the local, federal, and/or international policy and legal space.

This research uses the gender binary of "women" and "men," and therefore does not address the issues exclusively facing individuals who identify as nonbinary or gender nonconforming. One of the primary reasons for this focus is that most of the data collected on gender in the labour market and in education is based on the binary, particularly in the Gulf Cooperation Council (GCC), of which the UAE is a member. Of course, this lack of disaggregated data is an issue in and of itself, and the focus on the gender binary also often results in essentialism and further discrimination. As the AI Now Institute explained in its 2018 report when referring specifically to STEM (science, technology, engineering, and mathematics) pipeline research, "the persistent focus on gender as a binary often results in treating it as a biologically essential category that maps to certain attributes," which then leads to the implication that "the problem is one that resides within women's individual psychology…as opposed to an issue with the institutions and their cultures (Myers West et al. 2019)." Although it's beyond the scope of this research, it's important that researchers in this space work to broaden the perception of "gender" in the field of AI to address particular issues faced by individuals and groups who are not included in the gender binary.

By examining whether or not women's inclusion in AI is fundamental to the future success of nations in the 4IR, the hope is that this research may lead to further research regarding several important questions related to AI, including *why* nations that have national AI strategies to-date have not made women's inclusion (and gender diversity and equality more broadly) a strategic priority and *how* they should do so. As well as examining further how achieving gender equality in AI will help prevent building technology that will do more harm than good by exacerbating social ills, such as racial bias and religious discrimination. These questions are outlined in the conclusion of this chapter.

2 The Current Status of Women in the 4IR

If we were to look at world history as an astronaut looks at Earth from space, humanity has arguably benefited and continues to benefit from the progress made during the First, Second, and Third Industrial Revolutions (Chou 2019). For example, poverty rates have declined, literacy rates have increased, health conditions have improved, and political freedom has spread (Roser 2016). Women, in particular, have broadly benefited from this global climb as women's agency has increased, leading to women's rights movements and feminism (the advocacy for the equality of the sexes). As historian Estelle Freedman wrote in her book, *No Turning Back: The History of Feminism and the Future of Women*, "Feminist politics have originated where capitalism, industrial growth, democratic theory, and socialist critiques converged…" (Freedman 2002). It is imperative to recognise however that not all women have benefited equally from this progress—with white, Western women benefiting the most and primarily at the expense of racialised women. Further, those benefits were neither linear nor all-encompassing. In fact, social, economic, and political equality continue to elude many women across the globe (Seals Allers 2018). These racial inequalities and inequities still exist, as evident in the Black Lives Matters movement, which gained renewed energy in 2020 following the tragic killings of not only George Floyd, but also Breonna Taylor, Regis Korchinski-Paquet, Nina Pop, and countless others.

In 2020, the WEF reported that across the four subindexes observed in its *Global Gender Gap* report, there still remain significant gender gaps. These gaps are measured by calculating a series of female-to-male ratios on a set of variables in each respective subindex (World Economic Forum 2020). The aggregate of these ratios is then converted into percentages representing a female-to-male gap. The gaps listed in the 2020 report are Political Empowerment (75%), Economic Participation and Opportunity (42%), Educational Attainment (3%), and Health and Survival (3%) (Ibid.). The report states that the global, aggregate gender gap will take at least 99.5 years to close if trends continue—an improvement from the 2018 report. In the Middle East and North Africa (MENA), that number jumps to 139.9 years compared to 54.4 years in Western Europe and 151.4 years in North America (Ibid.). The COVID-19 pandemic, however, is estimated to negatively impact these numbers due to the precarious nature of women's labour participation, school closures, economic instability, and more (UN Women, 2020).

Narrowing the focus further, Figs. 1 and 2 show the global gender gap scores and the four subindexes from 2006 and 2020 for each member state of the GCC. From these tables, a few trends are noticeable. First,

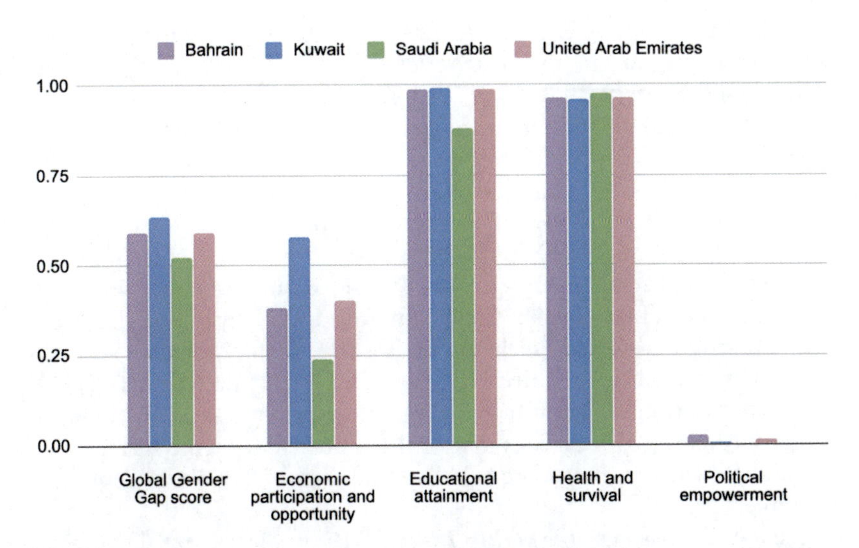

Fig. 1 World Economic Forum (WEF) Global Gender Gap Scores for GCC Member States (2006), sourced from the WEF's 2006 *Global Gender Gap* report

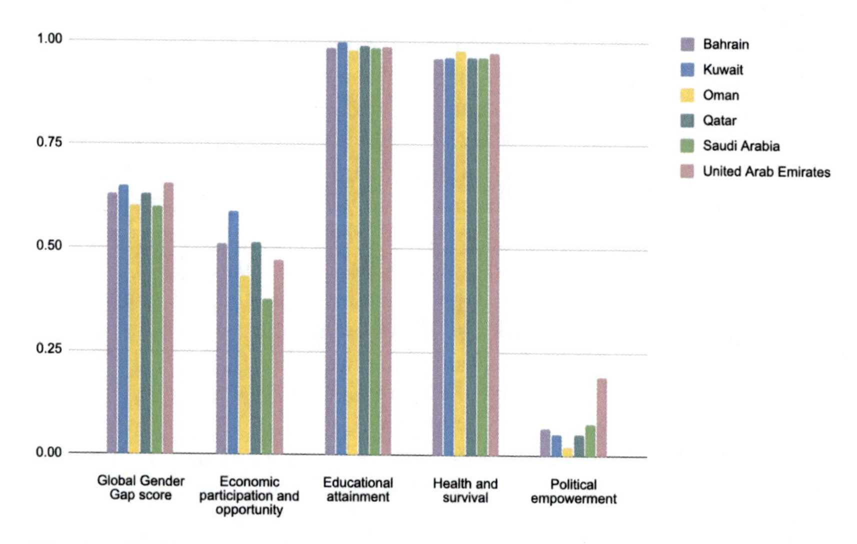

Fig. 2 World Economic Forum (WEF) Global Gender Gap Scores for GCC Member States (2020), sourced from the WEF's 2020 *Global Gender Gap* report

as evident in Fig. 1 vs Fig. 2, GCC member states have almost reached gender parity (1.0) in two categories: "Educational attainment" and "Health and Survival." However, they struggle to reach parity in "Economic participation and opportunity" and "Political Empowerment." Notably, the UAE has made progress since data was first collected in 2006 on "Political Empowerment," increasing their score from 0.015 in 2006 to a 0.191 in 2020 (Ibid.). This change is primarily the result of the UAE's Federal National Council (FNC) elections, which have increased its female membership every election year since 2006. Further, in 2019, UAE President H. H. Sheikh Khalifa bin Zayed Al Nahyan issued a directive requiring that women make up no less than 50% of member seats on the FNC by the next legislative cycle, an increase from 25% in 2018 (UAE National Elections Committee, 2019). Although the UAE's "Political Empowerment" score exceeds that of the other GCC member states, it's still far below the highest score held by Iceland at 0.701 (World Economic Forum 2020).

Finally, the *Gender Inequality Index* (GII) published by the United Nations Development Programme (UNDP), which ranks three aspects of human development—reproductive health, empowerment, and economic status—indicates that the overall trend of human development and gender equality is progressing upwards. In particular, out of the 160 countries included in the 2019 report, the GCC member states ranked: Qatar (45), Bahrain (47), UAE (26), Saudi Arabia (49), Oman (65), and Kuwait (53) (UNDP 2019).

Despite the progress made, however, there is still a lot of work to do, and with the arrival of the 4IR, there is both hope and fear for not only the future of women but also humanity as a whole. Within this new period of industrial progress, there exist many unknowns and important questions that we are not yet able to answer. As Njideka Harry, CEO of the Youth for Technology Foundation, stated in 2018:

> Will the 4IR be a revolution for women? It is critical to consider the impact of the Fourth Industrial Revolution on the gender gap. How will the accelerating pace of technological change affect what roles women can play in the economy, politics, and society? One question that stands out for me is how and whether the Fourth Industrial Revolution will worsen inequality, particularly for women. (Harry 2018)

Dr. Jill Rubery at the University of Manchester argues that in order to understand the gendered implications of AI on the labour market in particular, we have to acknowledge that new "technologies are not the only driver of change." We also have to take into account the "factors that have shaped and are reshaping current employment patterns by gender" (Rubery 2019). These questions become even more difficult to answer if the genuine inclusion of women in the 4IR remains elusive. "Genuine" inclusion means not only a notable increase in the number of women in STEM fields as well as decision-making roles across industries and governments, but also the retention and promotion of women in these roles over time.

As Caroline Criado Perez argues in her book, *Invisible Women: Data Bias in a World Designed for Men*, throughout history, the default for "human" was actually "male," and thus the impact of certain technologies (e.g. the design of seatbelts) on women was completely missed because of the absence of sex- and gender-segregated data and the lack of women involved in the development of these technologies (NPR 2019). We risk repeating similar mistakes in the 4IR if women, particularly women of colour who have historically been more severely underrepresented in

STEM, are not involved at all stages of the development, deployment, and governance of advanced technologies.

3 AI, the Technology Driving the 4IR

AI, the technology anticipated to profoundly impact all aspects of society and drive the 4IR forward, especially highlights women's lack of inclusion in STEM and why it matters. The concerns around AI are not only centred on how it will impact current gender inequalities, but also how it will impact already unequal and problematic political systems, information systems, labour markets, national and international security, health care, and education. All of which are underpinned by colonial patterns of global inequality, oppression, and discrimination that are already manifested in the field of AI—currently dominated by a handful of historically "powerful" and/or "wealthy" regions, such as North America, Europe, and East Asia (Mishra et al. 2019).

Interestingly, there is no shortage of female representation in AI products and services currently on the market and in the media. The most well-known examples are virtual assistants like Apple's *Siri*, social humanoid robots like Hanson Robotics *Sophia*, and AI news anchors like China's *Xin Xiaomeng*. The depiction of "female AI" has also been a common theme throughout movies and TV shows, with films such as *Metropolis* (1927), *Her* (2013), *Ex-Machina* (2014) and shows like *Westworld* (Sternberg 2018). Arguably, this is a reflection of not only the male-dominated industry creating these products, but also research that indicates people are more comfortable with a stereotypical female voice and likeness, particularly if the product is service-oriented. However, this also reflects unhealthy and harmful gendered social norms and gendered notions of work, as women were historically relegated to work in service-oriented roles (e.g. personal assistants, secretaries, etc.) that were considered inferior and low-paying (Nass et al. 1997).

Women's likeness, voice characteristics, and stereotypical behaviours are routinely utilised in AI products and services, far outweighing the actual number of women in the field. As the development of human-like interfaces, robots, and other technologies increases, it is important to understand the social impacts of product design and development decisions. This is particularly significant in regard to gender, as the ways in which we interact with these technologies and how we perceive them has the potential to further entrench bias and discrimination in the non-digital world (Eyssel and Hegel 2012). In feminist science and technology studies (STS), technology is neither "neutral" nor static, but in fact, "social relations (including gender relations) are materialised in tools and techniques" and the social meanings and use of technology change over time and context (Wajcman 2010). Therefore, the inclusion of women is imperative, as iterated by sociology professor Judy Wajcman in 2010:

> Empirical research on everything from the microwave oven, the telephone, and the contraceptive pill to robotics and software agents has clearly demonstrated that the marginalization of women from the technological community has a profound influence on the design, technical content, and use of artefacts. (Ibid.)

The AI Now Institute has also iterated Wajcman's point, stating, "The products of the AI industry already influence the lives of millions. Addressing diversity issues is therefore not just in the interest of the tech industry, but of everyone whose lives are affected by AI tools and services" (Myers West et al. 2019). Finally, Professor Bettina Buchel from the IMB Business School pointed to the specific impact on women:

> We are not only living in an age where women are being under-represented in many spheres of economic life, but technology could make this even worse...This is on the verge of being further reinforced by artificial intelligence, as current data being used to train machines to learn are often biased...Whenever you have a dataset of human decisions, it naturally includes bias. (Buchel 2018)

Further, research indicates that the lack of diversity in AI will lead to the development of AI tools and systems, such as facial recognition and "predictive policing" software that will replicate "patterns of racial and gender bias in ways that can deepen and justify historical inequality" (Myers West et al. 2019). For example, the current default of AI as "white" is already causing "representational harms," exacerbating racial bias and discrimination. One study on the "whiteness" of AI, referring not only to the predominant racial and ethnic make-up of the field, but also the visual presentation of AI technologies, argues:

> Portrayals of AI as White situate these machines in a power hierarchy above currently marginalised groups, such as people of color. These oppressed

> groups are therefore relegated to an even lower position in the hierarchy: below that of the machine...these portrayals could distort our perceptions of the risks and benefits of these machines. For example, they could frame the debate about AI's impact disproportionately around the opportunities and risks posed to White middle-class men. (Cave et al. 2020)

As evident above, the inclusion of racialised women and other marginalised groups in particular is not just a matter of improving workplace diversity in the field of AI (and STEM more broadly), but also a matter of preventing bias and discrimination from being built into AI systems from the beginning. Inclusion in the field is imperative to ensure that this technology doesn't affirm, sustain, and replicate the existing systems and structures that have proven harmful, oppressive, and discriminatory to whole groups of people.

Despite its importance, however, the inclusion of women in AI, particularly women of colour and women of low socioeconomic backgrounds, is abysmal. According to data gathered by the AI Now Institute, "women comprise 15% of AI research staff at Facebook and just 10% at Google...only 18% of authors at leading AI conferences are women and more than 80% of AI professors are male" (Ibid.; Mantha and Hudson 2018). In fact, only 22% of AI professionals worldwide identify as women (World Economic Forum 2018b). A report published by UNESCO in 2018 illustrated the gender gap in science more broadly. It found that the number of women researchers (defined as "professionals engaged in the conception or creation of new knowledge") was only 28.8% in the world and 39.8% for the Arab States, compared to 32.8% for North America and Western Europe (UNESCO 2018). For the GCC member states in particular, the number of women researchers is estimated to be: Kuwait (51.4%), Bahrain (39.0%), Qatar (31.4%), Oman (27.8%), Saudi Arabia (23.2%). Unfortunately, there was no listed data for the UAE in this report (Ibid.). Additional data shows that in some countries, such as the United States, the proportion of women working in computer science or engineering, fields from which AI draws, has fallen since its height in the 1980s–estimated to shrink the number of women in the field from 24% to 22% by 2025 (Simonite 2018; Accenture 2016).

Further, data collected by the WEF and LinkedIn indicates a significant gender gap in the AI industry and in the acquisition of AI skills. Although women and men have been acquiring AI skills "in tandem" since 2015, there is a persistent gap that corresponds with pre-existing gender gaps in the industry. More than half of the individuals with AI skills work in traditionally male-dominated industries like Software and IT Services (40%) and Education (19%), and women make up only 12% of that 59% share of AI talent (World Economic Forum 2018b). The data also shows that while both men and women are acquiring similar skills, such as machine learning, deep learning, computer vision, and Apache Spark, men are acquiring these emerging and more specialised skills in greater numbers. In parallel, women are acquiring skills such as data structures and information retrieval in greater numbers.

These trends correspond with data showing men are better represented in higher-paying and "more prestigious" occupational roles such as software engineer, business owner/founder, head of IT, head of engineering, and chief executive officer, and women are better represented in roles such as data analysts, researchers, and information managers. These trends demonstrate "a persistent structural gender gap among AI professionals. The gender gaps evident within the AI talent pool reflect both the broader gender gaps within specialisations in STEM studies; gender gaps across industries; and gender gaps in the acquisitions of emerging skills" (Ibid.).

These gender gaps are not the result of a simple "pipeline problem," however. As pipeline research shows, closing the gap requires more complicated and nuanced policies and actions than simply encouraging more girls to study STEM so that they eventually enter the workforce, particularly for fields such as engineering and computer science. Evidence shows that even when women do enter the STEM workforce, they are often pushed out or overlooked for leadership positions due to a variety of factors, such as sex segregation, gender socialisation, the motherhood penalty, lack of work-life balance, insufficient parental leave policies, pay inequity, toxic workplace culture, the role of stereotypes, and other structural barriers that stretch from K-12 education to the workplace (Myers West et al. 2019; Reinking and Martin 2018).

The skills and occupational gender gaps outlined above should be particularly worrying for governments hoping to lead in the 4IR due to the fact that, as outlined previously, "training, sustaining, and enabling an AI-capable talent pool" will be a key element determining national power. Therefore, it is surprising that addressing these particular issues is not made an explicit priority in any of the current, public national AI strategies. The section below will examine these strategies in more detail.

4 International and National AI Governance

This research focuses specifically on the inclusion of women (or gender more broadly) in national AI strategies because they are the foundational documents and policies released by national governments to signal their intentions and priorities regarding AI to domestic constituents and stakeholders, as well as international partners and adversaries. Historically, national strategies were a military tool but have over time been adopted by a range of stakeholders, such as civil society and political parties, and cover a wide array of issues, from women's economic empowerment to countering-violent extremism to climate change (van Eekelen 2010). It is important to examine these strategies because of the importance they play in the development of international norms and regulations regarding the development and deployment of AI.

It is important to note that it is difficult to compare and contrast any of the current national and/or regional AI strategies due to the fact that they vary immensely in approaches and policy focus (Dutton et al. 2018). This is an important caveat and highlights the lack of international norms and governance on the use of AI by governments and industry. Recent emphasis has been placed on the lack of international governance and norms regarding the militarisation of AI, with the failure of the UN Convention on Certain Conventional Weapons (UNCCW) to create a consensus on the regulation of lethal autonomous weapons. However, there is also a lack of international consensus around the broader development and application of AI, leaving private organisations like the IEEE, a professional engineers organisation, to develop codes, norms, and guidelines for ensuring "that every technologist prioritises ethical considerations in the design and development of autonomous and intelligent systems" (Rickli 2018).

More recently, there has been some movement on the international governance front in the form of international commitments and the formation of advisory bodies and guidelines to promote AI ethics and fairness principles. For example in 2018, the UN Secretary-General António Guterres announced the creation of the High-level Panel on Digital Cooperation tasked with contributing "to the broader public debate on the importance of cooperative and interdisciplinary approaches to ensure a safe and inclusive digital future for all taking into account relevant human rights norms." The UAE is represented on this 20-person panel by Mohammed Al Gergawi (UAE), Minister of Cabinet Affairs and the

Future (United Nations 2018). Additionally, in 2018 the International Study Group of Artificial Intelligence was created with support from France and Canada and the *G7 Charlevoix Common Vision for the Future of Artificial Intelligence* was launched to indicate commitments and/or actions towards international norms and/or governance regarding AI (Government of Canada 2019). The *Charlevoix* agreement does mention gender in several of the individual commitments listed, including number four, which states: "Support and involve women, underrepresented populations and marginalized individuals as creators, stakeholders, leaders and decision-makers at all stages of the development and implementation of AI applications" (Government of Canada 2019). More recently, the European Union (EU) released its *Ethics Guidelines for Trustworthy AI* in April 2019, developed by a High Level Expert Group on AI (Kung et al. 2020). It's important to note that many of the "international" efforts in creating ethical AI guidelines remain concentrated in and led by only a handful of regions (e.g. North America, Europe, and East Asia). Of course, this lack of regional inclusion means we risk creating irrelevant and ineffective solutions that fail to ensure AI doesn't harm historically marginalised communities and underrepresented countries that tend to bear the brunt of harms resulting from advances in technology (Hagerty et al. 2020). This trend is also evident in the national and/or regional AI strategies that have been developed to-date.

According to the second edition of CIFAR's report *Building an AI World*, there are 28 published national and/or regional AI strategies as of January 2020 (there were 18 in 2018). This chapter utilises the definition of "national strategy" developed by CIFAR in their 2018 report:

> ...an AI strategy is defined as a set of coordinated government policies that have a clear objective of maximizing the potential benefits and minimizing the potential costs of AI for the economy and society. The key word in this definition is coordinated because some countries have related AI policies in place that are uncoordinated. (Dutton et al. 2018)

Table 1 lists these different strategies and their policy areas as categorised by CIFAR. Each strategies' policy areas are assigned a "specificity" value based on the "specific policy measures" outlined and "allocated funding" received in the strategy. A policy area that is highly specific receives a value of four or five, medium specificity receives a value of two or three, and low specificity receives a value of one or zero (Kung et al.

Table 1 Overview of national and/or regional AI strategies or guiding documents

Country/region	*Title*	*Policy Areas (with specificity value)*
Canada	CIFAR Pan-Canadian Artificial Intelligence Strategy	• Research (5) • AI Talent (5) • Future of Work (0) • Industrial Policy (1) • Ethics (3) • Data and Digital Infrastructure (0) • AI in Government (0) • Inclusion (0)
China	A Next Generation Artificial Intelligence Development Plan	• Research (2) • AI Talent (2) • Future of Work (2) • Industrial Policy (3) • Ethics (3) • Data and Digital Infrastructure (3) • AI in Government (4) • Inclusion (0)
Czech Republic	National Artificial Intelligence Strategy of the Czech Republic	• Research (3) • AI Talent (5) • Future of Work (5) • Industrial Policy (5) • Ethics (4) • Data and Digital Infrastructure (5) • AI in Government (4) • Inclusion (4)

Country/region	*Title*	*Policy Areas (with specificity value)*
Denmark	National Strategy for Artificial Intelligence	• Research (2) • AI Talent (4) • Future of Work (1) • Industrial Policy (4) • Ethics (5) • Data and Digital Infrastructure (5) • AI in Government (4) • Inclusion (4)
Estonia	National AI Strategy 2019–2021	• Research (4) • AI Talent (5) • Future of Work (5) • Industrial Policy (5) • Ethics (4) • Data and Digital Infrastructure (3) • AI in Government (5) • Inclusion (0)
European Union	Coordinated Plan on AI	• Research (3) • AI Talent (4) • Future of Work (3) • Industrial Policy (5) • Ethics (5) • Data and Digital Infrastructure (5) • AI in Government (5) • Inclusion (4)

(continued)

Table 1 (continued)

Country/region	*Title*	*Policy Areas (with specificity value)*
Finland	Leading the way into the age of artificial intelligence	• Research (4) • AI Talent (4) • Future of Work (5) • Industrial Policy (5) • Ethics (5) • Data and Digital Infrastructure (5) • AI in Government (5) • Inclusion (2)
France	AI for Humanity: French Strategy for Artificial Intelligence	• Research (5) • AI Talent (5) • Future of Work (0) • Industrial Policy (5) • Ethics (5) • Data and Digital Infrastructure (5) • AI in Government (1) • Inclusion (2)
Germany	AI Made in Germany	• Research (5) • AI Talent (5) • Future of Work (5) • Industrial Policy (5) • Ethics (5) • Data and Digital Infrastructure (4) • AI in Government (1) • Inclusion (5)

Country/region	*Title*	*Policy Areas (with specificity value)*
India	National Strategy for Artificial Intelligence: #AIforAll	• Research (4) • AI Talent (5) • Future of Work (5) • Industrial Policy (5) • Ethics (5) • Data and Digital Infrastructure (5) • AI in Government (3) • Inclusion (1)
Japan	Artificial Intelligence Technology Strategy	• Research (2) • AI Talent (3) • Future of Work (0) • Industrial Policy (4) • Ethics (0) • Data and Digital Infrastructure (3) • AI in Government (0) • Inclusion (0)
Lithuania	Lithuanian Artificial Intelligence Strategy: A Vision of the Future	• Research (4) • AI Talent (5) • Future of Work (5) • Industrial Policy (2) • Ethics (5) • Data and Digital Infrastructure (4) • AI in Government (4) • Inclusion (2)

(continued)

Table 1 (continued)

Country/region	*Title*	*Policy Areas (with specificity value)*
Luxembourg	Artificial Intelligence: A Strategic Vision for Luxembourg	• Research (4) • AI Talent (3) • Future of Work (4) • Industrial Policy (5) • Ethics (4) • Data and Digital Infrastructure (4) • AI in Government (5) • Inclusion (2)
Malta	Malta: The Ultimate AI Launchpad—A Strategy and Vision for Artificial Intelligence in Malta 2030	• Research (4) • AI Talent (5) • Future of Work (5) • Industrial Policy (5) • Ethics (4) • Data and Digital Infrastructure (5) • AI in Government (5) • Inclusion (0)
Mexico	Towards an AI Strategy in Mexico: Harnessing the AI Revolution	• Research (3) • AI Talent (3) • Future of Work (1) • Industrial Policy (1) • Ethics (3) • Data and Digital Infrastructure (3) • AI in Government (2) • Inclusion (0)

Country/region	*Title*	*Policy Areas (with specificity value)*
Netherlands	Strategic Action Plan for Artificial Intelligence	• Research (4) • AI Talent (2) • Future of Work (5) • Industrial Policy (5) • Ethics (5) • Data and Digital Infrastructure (4) • AI in Government (5) • Inclusion (1)
Norway	The National Strategy for Artificial Intelligence	• Research (3) • AI Talent (3) • Future of Work (3) • Industrial Policy (4) • Ethics (4) • Data and Digital Infrastructure (5) • AI in Government (4) • Inclusion (2)
Portugal	AI Portugal 2020	• Research (5) • AI Talent (4) • Future of Work (5) • Industrial Policy (4) • Ethics (3) • Data and Digital Infrastructure (3) • AI in Government (4) • Inclusion (3)

(continued)

Table 1 (continued)

Country/region	*Title*	*Policy Areas (with specificity value)*
Qatar	Blueprint: National Artificial Intelligence Strategy for Qatar	• Research (2) • AI Talent (3) • Future of Work (5) • Industrial Policy (4) • Ethics (4) • Data and Digital Infrastructure (5) • AI in Government (0) • Inclusion (0)
Russia	National Strategy for the Development of Artificial Intelligence	• Research (2) • AI Talent (3) • Future of Work (2) • Industrial Policy (4) • Ethics (2) • Data and Digital Infrastructure (4) • AI in Government (3) • Inclusion (1)
Serbia	Strategy for the Development of Artificial Intelligence in the Republic of Serbia for the period 2020-2025	• Research (5) • AI Talent (3) • Future of Work (4) • Industrial Policy (4) • Ethics (5) • Data and Digital Infrastructure (4) • AI in Government (3) • Inclusion (5)

Country/region	*Title*	*Policy Areas (with specificity value)*
Singapore	AI Singapore	• Research (5) • AI Talent (5) • Future of Work (0) • Industrial Policy (5) • Ethics (0) • Data and Digital Infrastructure (0) • AI in Government (0) • Inclusion (0)
South Korea	Artificial Intelligence R&D Strategy	• Research (5) • AI Talent (5) • Future of Work (0) • Industrial Policy (5) • Ethics (0) • Data and Digital Infrastructure (0) • AI in Government (0) • Inclusion (0)
Sweden	National Approach to Artificial Intelligence	• Research (0) • AI Talent (3) • Future of Work (1) • Industrial Policy (3) • Ethics (3) • Data and Digital Infrastructure (3) • AI in Government (0) • Inclusion (0)

(continued)

Table 1 (continued)

Country/region	*Title*	*Policy Areas (with specificity value)*
Taiwan	Taiwan AI Action Plan	• Research (5) • AI Talent (5) • Future of Work (0) • Industrial Policy (5) • Ethics (0) • Data and Digital Infrastructure (4) • AI in Government (0) • Inclusion (0)
United Arab Emirates	UAE Strategy for Artificial Intelligence	• Research (0) • AI Talent (0) • Future of Work (0) • Industrial Policy (4) • Ethics (2) • Data and Digital Infrastructure (0) • AI in Government (2) • Inclusion (0)
United Kingdom	Industrial Strategy: Artificial Intelligence Sector Deal	• Research (5) • AI Talent (5) • Future of Work (4) • Industrial Policy (4) • Ethics (4) • Data and Digital Infrastructure (4) • AI in Government (4) • Inclusion (4)

Country/region	*Title*	*Policy Areas (with specificity value)*
United States	American AI Initiative	• Research (5) • AI Talent (4) • Future of Work (4) • Industrial Policy (2) • Ethics (2) • Data and Digital Infrastructure (5) • AI in Government (0) • Inclusion (3)

2020). It's generally understood that the more specificity assigned to a policy area, the more of a strategic priority that area is for the country and/or regional body.

The policy area most relevant to this chapter is "Inclusion and Social Well-Being" which is defined by CIFAR as, "Ensuring that AI is used to promote social and inclusive growth and that the AI community is inclusive of diverse backgrounds and perspectives" (Ibid.). To some extent, these policy areas are also relevant: "AI Talent," "Future of Work," and "Ethics (Ibid.)."

To understand better the policy area of "Inclusion," Fig. 3 illustrates this policy area and how many national and/or regional strategies received each specificity value. A significant observation from Fig. 3 is that out of the 28 national/regional AI strategies included, only six were assigned a specificity value of four or five, while 12 were assigned a zero. This means that only the Czech Republic, Denmark, the European Union, the United Kingdom, Germany, and Serbia have provided funding and/or significant strategic attention to the policy area, "Inclusion and Social Well-Being" (Ibid.).

Despite the lack of attention to "Inclusion" for a majority of the national/regional strategies, 16 received a specificity value of four or five in regard to the policy area "Ethics." This priority is defined as "The

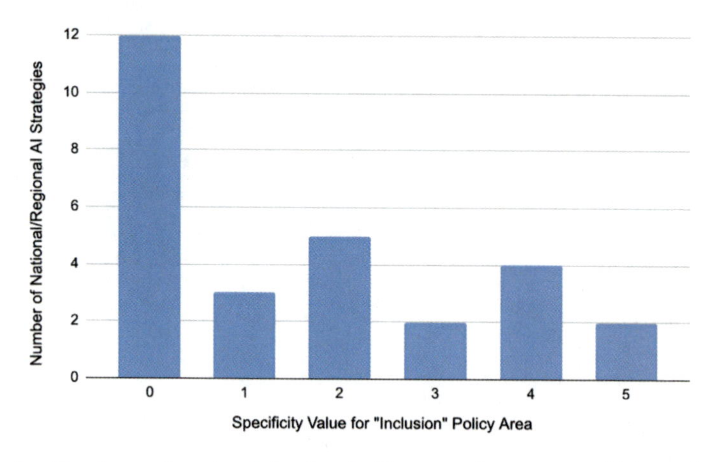

Fig. 3 Number of national/regional AI strategies per specificity value for "Inclusion" policy area, sourced from CIFAR's *Building an AI World* (Kung et al. 2020)

creation of a council, committee, or task force to create standards or regulations for the ethical use and development of AI. This area also includes specific funding for research or pilot programs to create explainable and transparent AI" (Ibid.). The details regarding this priority predominantly focus on the safe use of AI, regulatory actions, as well as the creation of platforms, councils, and/or research bodies to debate and develop best practices and/or guidelines. After digging further, however, this research could find no explicit mention of women's inclusion in these strategic priorities to-date, despite the evidence outlined above that clearly indicates the gender gap in AI talent.

Two priorities in which issues pertaining to gender may be implicitly included are "Skills and the Future of Work," defined as, "Initiatives to help students and the overall labour force develop skills for the future of work, such as investments in STEM education, digital skills, or lifelong learning," and "AI Talent Development," defined as, "Funding to attract, retain, and train domestic or international AI talent, including funding for chairs and fellowships or the creation of AI-specific Master and PhD programs (Ibid.)."

An important note is that women's inclusion, or gender more broadly, may not be listed explicitly in these strategies if the corresponding national government has already established a comprehensive and encompassing inclusion strategy throughout all of its policy and/or legislative work. This could be the case in more egalitarian nations, such as Denmark and Sweden, that have been more successful than others in undertaking gender mainstreaming and implementing policies such as quotas, pay equity, and anti-discrimination laws (Caglar 2013). However, these are also the countries struggling the most with increasing the number of women in STEM fields. Recent trends indicate a "gender-equality paradox" in which "countries with high levels of gender equality have some of the largest STEM gaps in secondary and tertiary education" and thus fewer women entering the STEM workforce (Stoet and Geary 2018). In fact, data from UNESCO on female STEM researchers shows that of the European countries listed (41), Denmark ranks 32nd (only 32.5% of researchers are women) and Sweden ranks 30th (only 33.7% of researchers are women). It can be assumed, due to broader trends in AI, that the number of female researchers with a specialisation in AI is even lower. Therefore, it would be in these governments' interest to make the inclusion of women an explicit priority in their national AI strategies to help close this gap.

5 Case Study: An Examination of the United Arab Emirates (UAE)

In order to examine the research question further, it is helpful to zero in on a specific national context. Particularly one in which AI is a significant national strategic priority and the "empowerment of women" is also considered a national strategic priority. The UAE has claimed both over the last few years.

In 2015, the UAE government launched its *National Strategy for Empowerment of Emirati Women in the UAE*, focused on the years 2015–2021. This national strategy "aims to provide a framework for all federal and local government entities, private sector, as well as social organizations, to set plans that will provide a decent living for women and make them creative in all sustainable and developmental fields" (Government of the United Arab Emirates 2018a). After the strategy's announcement, the UAE Gender Balance Council was formed to help boost "gender balance" in the public and private sector and ensure the strategy's success (Gulf News 2016). Although vague, this national strategy solidifies actions the Emirati government has undertaken for several years to "empower" its female citizens.

The strategy above was followed by the launch of the *UAE Strategy for Artificial Intelligence* and the creation of the first-ever Minister of State for Artificial Intelligence, a position filled by H. E. Omar Al Olama, in 2017 (Government of the United Arab Emirates 2018b). This strategy was revamped in 2018 to be the UAE *AI Strategy 2031* and will be implemented by the newly formed Council of Artificial Intelligence and Digital Transactions (Government of the United Arab Emirates 2017). The Emirati government has clearly stated that AI is a significant priority, as Sheikh Mohammed bin Rashid Al Maktoum, the Vice President and Prime Minister of Dubai stated in 2019, "We have launched a national strategy for artificial intelligence to make it an integral part of our business, our lives, and our government services" (Gulf News 2019).

There are several other reasons why the UAE is an important case study. First, the UAE is one of only two member states of the GCC to release a coordinated national AI strategy and thus, it has the potential to influence the rest of the GCC's policies on AI, including Saudi Arabia which has also begun to invest heavily in AI through its Sovereign Wealth Fund (PwC 2018). Second, the UAE has significant demographic and economic issues that may only be exacerbated by advances in AI and reflect issues in the larger GCC. This includes a growing population under the age of 25 (estimated to be 31% in the UAE), increasing unemployment among nationals due to traditional preferences for the public sector over the private sector, a downturn in oil prices, and an education system that is struggling to adequately prepare students for the labour market (Ashkar and Anderson 2016). These issues may be worsened if reskilling and job displacement estimates turn out to be true. For example, the WEF estimates that 54% of employees will "require significant reskilling and upskilling" by 2022 as companies begin to automate and augment more tasks and processes (World Economic Forum 2018a). Another significant issue pertaining to the labour market and economic health is the fact that women's economic empowerment, a stated priority of the UAE government, may be hampered by AI. In fact, the WEF estimates that 57% of the jobs displaced by advances in AI will belong to women (Ibid.).

Third, AI has become a significant national security issue, particularly in the GCC and broader MENA region. Much has been said in international and national security circles regarding the impact of AI on traditional warfare (e.g. the impact of lethal autonomous weapons), but it is also important to note the vulnerabilities countries face as they become more digitised. As Omar bin Sultan Al Olama, Minister of State for Artificial Intelligence, stated in 2019:

> The more we become digitalised, the more we are vulnerable to hackers…It is very easy today for a nation to be attacked through hacking into its defence system, unlike before when it required physical invasion…It is important for nations to become AI exporters through innovations so as to reduce the risks of cyber attacks from imported technologies by hackers. (Sebugwaawo 2019)

Finally, due to the fact that AI is integral to the future of the UAE government and will be utilised extensively in its decision-making, understanding how the UAE prioritises issues such as the inclusion of women, as well as diversity and inclusion on the basis of race, ethnicity, religion, and immigration status, is critical to understanding the potential human rights abuses that may result from the deployment of AI (PwC 2018).

6 A Detailed Look at the UAE's National AI Strategy

As mentioned above, the UAE released its revamped *UAE AI Strategy 2031* in 2018, and it is not surprising that the UAE was one of the first countries to develop a coordinated and comprehensive national AI strategy (Government of the United Arab Emirates 2017). It is estimated that by 2031, AI could provide an additional AED 333bn in growth to the UAE's economy, and help transform it into a knowledge-based economy that is less volatile and more sustainable than one based on natural resources (Government of the United Arab Emirates 2019b).

For several decades, the UAE has sought to transform itself from an oil-based economy to a knowledge-based economy. Following its independence in 1971, the UAE utilised its oil-wealth to fund new industries and diversify its economy. Since the early 1990s, the government has sought to encourage innovation by "establishing Free Zones, setting up financial centres, and supporting education." After the 2008 global financial crisis, which particularly affected Dubai, the UAE launched the 2020 vision in which "innovation and entrepreneurship" became strategic priorities for the UAE's future economy (AbouHana 2017). It is within this context that the UAE's national AI strategy was developed. This strategy is an extension of the government's long-term strategy to not only transform its domestic economy, but to also be a global leader in innovation and technology.

According to the UAE government's website, the AI strategy's main objectives are to make the government more efficient, to create new economic opportunities, and to lead the world in AI investments. The strategy covers nine different sectors, including:

- Transportation to reduce accidents and cut operational costs
- Health care to minimise chronic and dangerous diseases
- Space to help conduct accurate experiments and reduce the rate of costly mistakes
- Renewable energy to manage facilities
- Water to conduct analysis and studies to provide water sources
- Technology to increase productivity and help with general spending
- Education to cut costs and enhance desire for education
- Environment to increase forestation rate

- Traffic to reduce accidents and traffic jams and create more effective traffic policies (Government of the United Arab Emirates 2017)
- The strategy also has five overarching themes, including:
- The formation of the UAE AI Council
- The creation of workshops, programmes, initiatives, and field visits to government bodies
- The development of the capabilities and skills of all staff operating in the field of technology and organise training courses for government officials
- The provision of all services via AI and the full integration of AI into medical and security services
- The launch of a leadership strategy and issue a government law on the safe use of AI (Ibid.)

Other components of the UAE's national AI strategy include the National Program for Artificial Intelligence, otherwise known as Building a Responsive Artificial Intelligence Nation (BRAIN), which is a "comprehensive and consolidated compilation of resources that highlight the advances in AI and Robotics" (Government of the United Arab Emirates 2019b). The UAE Council for Artificial Intelligence was also formed in March of 2018 and is "tasked with proposing policies to create an AI-friendly ecosystem, encourage advanced research in the sector and promote collaboration between the public and private sectors, including international institutions to accelerate the adoption of AI" (Ibid.). The UAE has also launched several educational and awards initiatives directed at Emirati nationals to increase and support local AI talent and innovation. These include the Artificial Intelligence Internship Program in partnership with Dell Technologies, the Artificial Intelligence Summer Camp, and the "AI and Robotics Award for Good" (Ibid.). The UAE government has also hosted several AI-related global summits, conferences, and events, including the AI Everything Summit, which is a "culmination of the UAE and the world's mission to promote initiatives, collaborations, partnerships and breakthroughs in the field of AI, and to foster positive impact on governments, businesses, social enterprises and humankind" (AI-everything 2019). Finally, the government most recently launched the Think AI Initiative in partnership with Ernst and Young (EY) for the purpose of "accelerating the integration of artificial intelligence in strategic sectors such as infrastructure, governance, and legislation." This initiative will include a series of discussions with a variety of stakeholders to debate certain issues pertaining to AI, including "regulation and governance" and developing "national standards" for AI (GIP Digital Watch observatory 2019).

From what has been made public about the UAE's national AI strategy, it appears to adhere to most of the objectives that a national strategy should ideally include, such as reflecting a "consensus in society about national interests and threats," serving as "the basis for more detailed sectoral documents," and acting as "a tool of public diplomacy" (Ibid.). The latter being the most notable, as the UAE has utilised this national AI strategy to signal both domestically and internationally its intention to be a global leader in AI. However, there is no explicit mention of women's inclusion in this strategy, and in fact, certain anecdotal evidence indicates that the UAE still has yet to make this a strategic priority. For example, out of the 16-member UAE Council for Artificial Intelligence, only three members are women: H.E. Dr. Rawda Saeed Al Saadi, Director General of Smart Solutions and Services Authority; H.E. Aisha Bin Bishr, Director-General of the Smart Dubai Office; and H.E. Ohood Al Shuhail, Director General of Ajman Digital Government (Ibid.). Another example is that out of the 198 speakers listed on the 2019 AI Everything Summit's website, only 28 were women (~ 15%). Out of those 28 women, it appears that only two were Emirati nationals; H.E. Aisha Bin Bishr, Director-General of the Smart Dubai Office; and Hessa Al Baloosh, Director of Smart Services Department, Smart Dubai Government Est.

7 The Importance of Women's Inclusion in the UAE's National AI Strategy

By focusing on one of the key elements of national power necessary in the 4IR, namely, "Training, sustaining, and enabling an AI-capable talent pool," we can see why the inclusion of women is so vital to the security, success, and prominence of the UAE in the 4IR (Horowitz et al. 2018).

In order to transform its economy to a knowledge-based economy and also become a global leader in AI, the UAE has made it a priority to shift away from relying on foreign labour (an estimated 95% of the workforce) towards relying on national labour, a process dubbed "Emiratisation" (Government of the United Arab Emirates 2019a). This strategy will, however, be unsuccessful if its female population is not included (Gallant and Pounder 2008). Fostering AI talent through education and reskilling programmes of all nationals will also grow in importance as the economy of the future takes shape (Buttorff et al. 2018). In fact, an estimated 47% of the jobs in the UAE are estimated to be susceptible to automation, and the WEF estimates that 21% of the core skills required in 2015 for employment in the broader GCC will change by 2020 (World Economic Forum 2017). This will likely increase the unemployment rate for Emirati nationals, which has risen since 2015, especially among women (The World Bank 2019b).

To support the desired shift to a knowledge-based economy and to become a leader in AI, the government has also focused on developing a "first-rate education system," launching its National Strategy for Higher Education in 2017. According to 2017 education statistics from the UAE Ministry of Education, Emirati women currently outnumber Emirati men in government higher education institutions (~ 74% vs. ~ 26%), although there is no major gender gap at private higher education institutions. When it comes to certain STEM fields, gender gaps also become apparent. In "Engineering" women make up 44% of the enrolled students, in "Sciences" women make up 95%, and in "Information Technology" women make up 70%. Interestingly, at the private higher education institutions, the gender gaps change. In "Engineering" women make up 53% of the enrolled students, in "Sciences" women make up 80%, and in "Information Technology" women make up 45% (UAE Ministry of Education 2018). This difference may be a result of access to certain programmes and/or disciplines in private vs. public universities (Ghazal Aswad et al. 2011). Finally, in aggregate, women comprise 49% of "Engineering" students, 93% of "Sciences" students, and 64% of "Information Technology" students.

With such a high percentage of female students graduating with STEM degrees, it would be safe to assume that these women would enter the workforce in relatively high numbers. That does not appear to be the case, however. According to data from 2017, the labour force participation rate for Emirati women is 32.4%, only a small improvement from 28.2% in 2008 (UAE Federal Competitiveness and Statistics Authority 2017). That rate is below the estimated global average of 48% (The World Bank 2019a). When it comes to women in STEM fields, particularly AI, that rate is estimated to be lower, although reliable statistics are not yet publically available for the UAE. Perhaps an indication is that, as previously mentioned, the 16-member UAE Council for Artificial Intelligence has only three women members. Research also shows that even if women do enter the STEM workforce in the UAE, there are several factors that often force them to leave, such as work-family balance constraints, lack of sufficient parental leave benefits, access to childcare, family pressure, discrimination, and low job satisfaction (The Economist Intelligence Unit 2014). It is this gap in the labour market that should be particularly disconcerting to the UAE government, and thus one of the reasons women's inclusion should be an explicit priority in its AI strategy.

Not only is the inclusion of women imperative to the success of the UAE's future economy and its AI ambitions, it is also important to the success of AI technologies developed and deployed by the UAE. If these technologies are being developed predominantly by men, as is currently the case, there is the potential for algorithms to entrench existing bias and inequality because they are being fed biased data—as discussed in previous sections of this chapter. This issue was exemplified by the AI recruiting system utilised by Amazon since 2015, which showed a profound gender bias due to the fact that the system was trained on data from previous applicants over a 10-year period, who were mostly men (Dastin 2018). The UAE suffers from particular social ills that AI may exacerbate, such as gender inequality, religious discrimination, political oppression, wealth inequality, and more. In order to avoid these issues, Ann Cairns, the Vice Chairman of Mastercard, argues that "[we need] more women working in the technology industry, writing algorithms and feeding into product development" (Cairns 2019). Women's inclusion will "change how we imagine and develop technology, and how it sounds and looks" (Ibid.). Therefore, women's inclusion is also a matter of importance for the ethical and successful development of AI technologies.

Further, if the UAE wants to be a leader in AI, it needs to take issues regarding women's inclusion at all stages of development, deployment, and governance of AI seriously. By making women's inclusion an explicit priority in its national AI strategy, the UAE would not only be providing an important policy guide domestically, but would also signal its commitment to women's inclusion in this space internationally. This is a signal that has yet to be clearly and eagerly emitted by any of the other nations hoping to lead this field.

8 How Should Women Be Included in National AI Strategies?

Answering this question fully goes beyond the scope of this chapter. However, it is an important question to examine briefly. In 2018, the

World Wide Web Foundation released a report providing clear recommendations on actions that governments can take to ensure the inclusion of women in AI at many different levels (Avila et al. 2018). The Foundation's recommendations include three distinct proposals:

1. Countries need to take proactive steps towards the inclusion of women in the coding and the design of machine learning and AI technologies
2. States need to implement industry guidelines to protect women from discriminatory algorithms and embrace openness and transparency for AI
3. Countries must assess the economic, political, and social effects of AI and machine learning technologies on the lives of women (Ibid.)

The Foundation's report, however, does not address how governments can make women's inclusion an explicit priority in their coordinated national AI strategies. A tool that may be used in order to do this is gender mainstreaming. Gender mainstreaming is defined by the United Nations as "the deliberate consideration of gender in all stages of program and policy planning, implementation and evaluation, with a view to incorporate the impact of gender at all levels of decision-making," and may be an effective baseline strategy that can serve as a defined priority for governments to include in their national AI strategies (UNESCO 2017). In practice, gender mainstreaming can take many forms and has been a relatively consistent and accepted strategy by national governments and international organisations for the "promotion of gender equality" since 1995 (United Nations 2011). Despite criticism of gender mainstreaming and its effectiveness, it remains a vital baseline for government action towards gender equality by including gendered perspectives in policies and/or agendas from the very beginning of their development (Ibid.). Therefore, gender mainstreaming in national AI strategies could be an important initial tool used by national governments to increase gender inclusion and gender equality in this emerging field. Of course, there are several other questions that must be asked when considering implementing gender mainstreaming in this way, including: What would the incorporation of gender mainstreaming in national AI strategies look like? Is it the best tool for promoting the inclusion of women and other underrepresented groups in AI? What are the mechanisms or indicators by which we can measure or track progress? Answering these questions falls outside of the scope of this chapter, however.

9 Conclusion

In conclusion, this research found that women's inclusion in the development, deployment, and governance of AI is extremely important for nations in the 4IR and therefore should be included in national AI strategies. This is for a variety of reasons which can be grouped into three broad themes, outlined below:

- Social—AI has the potential to exacerbate and entrench existing bias and inequality, particularly gender and racial inequality
- Security—Due to the fact that AI as a general-purpose technology, it has the potential to impact every aspect of society and this can have a major impact on national and international security
- Economic—As governments jockey for power in the 4IR, they need a workforce with the skills and abilities to match labour market demands, this includes AI skills and expertise

For this research, I focused particularly on the economic theme and the importance of achieving gender equality in the field of AI by including women in strategies to "train, sustain, and enable an AI-capable talent pool," one of the key elements of national power in the 4IR listed by Scharre and Horowtiz (Horowitz et al. 2018). I examined the research questions from a qualitative approach utilising academic research from a variety of fields, such as including security, business, science, women and gender studies, as well as relevant news articles and data from government and intergovernmental organisations, to understand the current status of women in AI.

In the first section of this chapter, I examined more broadly the status of women in the 4IR and why women's inclusion is important. Then, I examined the current state of national AI strategies and their priorities as categorised by the Canadian Institute for Advanced Research (CIFAR), and finally, proceeded to examine the United Arab Emirates (UAE) as a specific case study in order to further contextualise this research (Dutton et al. 2018). The use of the UAE as a case study, or the use of a case study at all, is important due to the fact that to-date, national AI strategies are quite varied and difficult to study from a broad perspective. Therefore, examining a specific case study offers a better analysis of the research questions. In particular, the UAE is a useful case study because it has not only made AI a significant national strategic priority—evident in the creation of its *UAE Strategy for Artificial Intelligence* and the world's first Minister of State for Artificial Intelligence in 2017—but it has also indicated that the "empowerment of women" is a national strategic priority with the launch of the *National Strategy for the Empowerment of Emirati Women* in 2015.

This research raises several additional questions and areas for further research, including:

1. What role could the UAE play in influencing and/or informing the development of AI strategies by the other GCC members? Particularly in regard to women's inclusion and gender equality?
2. Why has none of the countries with national AI strategies exclusively made women's inclusion, or gender more broadly, an explicit strategic priority?
3. What role do international norms play in the setting of strategic priorities in regard to AI? How does international norm diffusion operate in AI policy and/or governance?
4. How do the specific strategic priorities outlined in current national AI strategies interact with one another? What takes precedence? How are the factors that influence how these priorities are set (e.g. corporate interests, security interests, etc.)?
5. What role do various identity factors, such as race, religion, ethnicity, disability, and indigeneity play in the exclusion and/or inclusion of certain people in the AI field?

In order to conduct the above research, however, more data is needed not only from the UAE and the GCC, but also globally. In particular, data that captures the demographics and intersectional identities of those studying AI-related fields, acquiring AI-skills, and working in AI is necessary to better understand the current state of the field and to develop effective policy interventions. This data is necessary to better understand the current state of the field and to develop effective policy interventions. Finally, a more detailed, consistent, and up-to-date database that captures national AI strategies, including the priorities, actions, progress, funding, etc., would prove incredibly useful for future research into AI policy and governance.

References

AbouHana, Mona. (2017). Innovation in the UAE: From First Foundations to 'Beyond Oil. PWC. Retrieved from: https://www.pwc.com/m1/en/media-centre/articles/innovation-in-the-uae-from-first-foundations-to-beyond-oil.html.

Accenture (2016). Cracking the gender code. Accenture and Girls Who Code. Retrieved from: https://www.accenture.com/us-en/cracking-the-gender-code?src=JB-11540.

AI-everything (2019). About Ai Everything. Ai Everything Summit. Retrieved from: https://ai-everything.com/about-ai-everything/.

Allers, Kimberly Seals. (2018). Rethinking work-life balance for women of color: And how women got it in the first place. Slate. Retrieved from: https://slate.com/human-interest/2018/03/for-women-of-color-work-life-balance-is-a-different-kind-of-problem.html.

Ashkar, Hani and Anderson, Stephen. (2016). Middle East Megatrends: Transforming our region. PWC Middle East. Retrieved from: https://www.pwc.com/m1/en/issues/megatrends.html.

Aswad, Noor Ghazal; Vidican, Georgeta; and Samulewicz, Diana. (2011). Creating a knowledge-based economy in the United Arab Emirates: Realizing the unfulfilled potential of women in the science, technology, and engineering fields. European Journal of Engineering Education, 36(6), 559–570.

Avila, Renata; Brandusescu, Ana; Freuler, Juan Ortiz; and Thakur, Dhanaraj. (2018). Artificial Intelligence: open questions about gender inclusion. World Wide Web Foundation. Retrieved from: http://webfoundation.org/docs/2018/06/AI-Gender.pdf.

Buchel, Bettina. (2018). Artificial intelligence could reinforce society's gender equity problems. The Conversation. Retrieved from: https://theconversation.com/artificial-intelligence-could-reinforce-societys-gender-equality-problems-92631.

Buttorff, Gail; Welborne, Bozena; and al-Lawati, Nawra. (2018). Measuring Female Labor Force Participation in the GCC. Baker Institute for Public Policy, Rice University, 1–6.

Caglar, Gulay. (2013). Gender Mainstreaming. Politics and Gender, 9(3), 336–344.

Cairns, Ann. (2019). Why AI is failing the next generation of women. World Economic Forum. Retrieved from: https://www.weforum.org/agenda/2019/01/ai-artificial-intelligence-failing-next-generation-women-bias/.

Cave, Stephen; Dihal, Kanta. (2020). The Whiteness of AI. Philosophy and Technology. Retrieved from: https://link.springer.com/article/10.1007/s13347-020-00415-6.

Chou, Shuo-Yan. (2019). The Fourth Industrial Revolution: Digital Fusion with the Internet of Things. Journal of International Affairs, 72(1), 107–120.

Daemmrich, Arthur. (November 04, 2017). Invention, Innovation Systems, and the Industrial Revolution. Technology and Innovation. 18, 259–260.

Dafoe, Allan. (2018). AI Governance: A Research Agenda. Future of Humanity Institute, University of Oxford. Retrieved from: https://www.fhi.ox.ac.uk/wp-content/uploads/GovAIAgenda.pdf.

Dastin, Jeffrey. (2018). Amazon scraps a secret AI recruiting tool that showed bias against women. Reuters. Retrieved from: https://www.reuters.com/article/us-amazon-com-jobs-automation-insight/amazon-scraps-secret-ai-recruiting-tool-that-showed-bias-against-women-idUSKCN1MK08G.

Davis, Nicholas and Philbeck, Thomas. (2019). The Fourth Industrial Revolution: Shaping of a New Era. Journal of International Affairs, 72(1), 17–22.

Dutton, Tim; Barron, Brent; and Boskovic, Gaga. (2018). Building an AI World: Report on National and Regional AI Strategies. Canadian Institute for Advanced Research (CIFAR), 1–30.

Eekelen, William F. van. (2010). The Definition of a National Strategic Concept. Geneva Centre for the Democratic Control of Armed Forces and Asian Study Centre for Peace & Conflict Transformation, 1–24.

Eyssel, Friederike and Hegel, Frank. (2012). (S)he's Got the Look: Gender Stereotyping of Robots. Journal of Applied Social Psychology, 42(9), 2213–2230.

Freedman, Estelle. (2002). No Turning Back: The History of Feminism and the Future of Women. New York: The Random House Publishing Group.

Gallant, Monica and Pounder, James S. (2008). The employment of female nationals in the United Arab Emirates (UAE). Education, Business, and Society: Contemporary Middle Eastern Studies, 1(1), 26–33.

GIP Digital Watch observatory (2019). UAE government launches Think AI initiative. Geneva Internet Platform. The Geneva Internet Platform Digital Watch observatory. Retrieved from: https://dig.watch/updates/uae-government-launches-think-ai-initiative.

Government of Canada (2018). Canada-France Statement on Artificial Intelligence. Government of Canada. Retrieved from: https://international.gc.ca/world-monde/international_relations-relations_internationales/europe/2018-06-07-france_ai-ia_france.aspx?lang=eng.

Government of Canada (2019). Charlevoix common vision for the future of artificial intelligence. Government of Canada. Retrieved from: https://intern

ational.gc.ca/world-monde/international_relations-relations_internationales/g7/documents/2018-06-09-artificial-intelligence-artificielle.aspx?lang=eng.
Government of the United Arab Emirates (2017). UAE Strategy for Artificial Intelligence," Government of the United Arab Emirates. Retrieved from: http://www.uaeai.ae/en/.
Government of the United Arab Emirates (2018a). National Strategy for the Empowerment of Emirati Women. Government of the United Arab Emirates. Retrieved from: https://government.ae/en/about-the-uae/strategies-initiatives-and-awards/federal-governments-strategies-and-plans/national-strategy-for-empowerment-of-emirati-women.
Government of the United Arab Emirates (2018b). UAE Strategy for Artificial Intelligence.
Government of the United Arab Emirates (2019a). Emiratisation. Government of the United Arab Emirates. Retrieved from: https://www.government.ae/en/information-and-services/jobs/vision-2021-and-emiratisation/emiratisation.
Government of the United Arab Emirates (2019b). National Program for Artificial Intelligence. World Leaders in AI by 2031. Government of the United Arab Emirates. Retrieved from: https://ai.gov.ae/about-us/.
Gulf News (2016). UAE Gender Balance Council to help boost equality. Gulf News. Retrieved from: https://gulfnews.com/uae/government/uae-gender-balance-council-to-help-boost-equality-1.1816276.
Gulf News (2019). Shaikh Mohammad Bin Rashid chairs Cabinet meeting for AI strategy 2031. Gulf News. Retrieved from: https://gulfnews.com/uae/government/shaikh-mohammad-bin-rashid-chairs-cabinet-meeting-for-ai-strategy-2031-1.1555856589838.
Hagerty, Alexa; Rubinov, Igor. (2020). Global AI Ethics: A Review of the Social Impacts and Ethical Implications of Artificial Intelligence. Retrieved from: https://arxiv.org/abs/1907.07892.
Harry, Njideka. (2018). Will the Fourth Industrial Revolution be a revolution for women? World Economic Forum. Retrieved from: https://www.weforum.org/agenda/2018/01/gender-inequality-and-the-fourth-industrial-revolution/..
Horowitz, Michael; Gregory, Allen C.; Kania, Elsa B.; and Scharre, Paul. (2018). Strategic Competition in the Era of Artificial Intelligence. Center for a New American Security (CNAS), 1–27.
Jackson, Chris. (2017). Perils of Perception: Global Impact of Development Aid. Ipsos. Retrieved from: https://www.ipsos.com/en/global-perceptions-development-progress-perils-perceptions-research.
Kung, Johnny; Boskovic, Gaga; Stix, Charlotte. (2020). Building an AI World: Report on National and Regional AI Strategies Second Edition. CIFAR. Retrieved from: https://www.cifar.ca/ai/building-an-ai-world-second-edition.

Mantha, Yoan and Hudson, Simon. (2018). Estimating the Gender Ratio of AI Researchers Around the World. Element AI. Retrieved from: https://medium.com/element-ai-research-lab/estimating-the-gender-ratio-of-ai-researchers-around-the-world-81d2b8dbe9c3.

Ministry of Foreign Affairs (2018). Charlevoix Common Vision for the Future of Artificial Intelligence. G7. Retrieved from: https://www.mofa.go.jp/files/000373837.pdf.

Mishra, Saurabh. (2019). Artificial Intelligence Index: 2019 Annual Report. Stanford Human-Centered Artificial Intelligence. Retrieved from: https://hai.stanford.edu/sites/default/files/ai_index_2019_report.pdf.

Nass, Clifford; Moon, Youngme; and Green, Nancy. (1997). Are Machines Gender Neutral? Gender-Stereotypic Responses to Computers With Voices. Journal of Applied Social Psychology, 27(10), 864–876.

National Public Radio (NPR) (2019). Caroline Criado-Perez on data bias and 'Invisible Women. NPR. Retrieved from: https://www.npr.org/2019/03/17/704209639/caroline-criado-perez-on-data-bias-and-invisible-women.

PwC Middle East (2018). US$320 billion by 2030? The potential impact of AI in the Middle East. PwC Middle East. Retrieved from: https://www.pwc.com/m1/en/publications/documents/economic-potential-ai-middle-east.pdf.

Reinking, Anni and Martin, Barbara. (2018). The Gender Gap in STEM Fields: Theories, Movements and Ideas to Engage Girls in STEM. Journal of New Approaches in Educational Research, 7(2), 148–153.

Rickli, Jean-Marc. (2018). The Economic, Security, and Military Implications of Artificial Intelligence for the Arab Gulf Countries. Emirates Diplomatic Academy, 1–13.

Roser, Max. (2016). Proof that life is getting better for humanity, in 5 charts. Vox Media. Retrieved from: https://www.vox.com/the-big-idea/2016/12/23/14062168/history-global-conditions-charts-life-span-poverty.

Rubery, Jill (2019). A Gender Lens on the Future of Work. Journal of International Affairs, 72(1), 91–106.

Scharre, Paul and Horowitz, Michael. (2018). Artificial Intelligence: What Every Policymaker Needs to Know. Center for a New American Security (CNAS). Retrieved from: https://www.cnas.org/publications/reports/artificial-intelligence-what-every-policymaker-needs-to-know.

Schwab, Klaus. (2016). The Fourth Industrial Revolution: what it means, how to respond. World Economic Forum. Retrieved from: https://www.weforum.org/agenda/2016/01/the-fourth-industrial-revolution-what-it-means-and-how-to-respond/.

Sebugwaawo, I. (2019). Nations more vulnerable to attacks in the digitised world: AI minister. Khaleej Times. Retrieved from: https://www.khaleejtimes.com/nation/abu-dhabi/nations-more-vulnerable-to-attacks-in-digitsed-world-ai-minister.

Simonite, Tom. (2018). AI is the future–but where are the women? Wired. Retrieved from: https://www.wired.com/story/artificial-intelligence-researchers-gender-imbalance/.

Sternberg, Irene. (2018). Female AI: The Intersection Between Gender and Contemporary Artificial Intelligence. Hackernoon. Retrieved from: https://hackernoon.com/female-ai-the-intersection-between-gender-and-contemporary-artificial-intelligence-6e098d10ea77.

Stoet, Gijsbert and Geary, David C. (2018). The Gender-Equality Paradox in Science, Technology, Engineering, and Mathematics Education. Association for Psychological Science, 29(4), 581–593.

The Economist Intelligence Unit (2014). UAE Economic Vision: Women in Science, Technology and Engineering. The Economist Intelligence Unit, 1–31.

The World Bank (2019a). Labor force participation rate, female (% of female population ages 15 +) (modeled ILO estimate). The World Bank. Retrieved from: https://data.worldbank.org/indicator/sl.tlf.cact.fe.zs?end=2017&start=1990.

The World Bank (2019b). Unemployment, female (% of female labor force) (modeled ILO estimate). World Bank. Retrieved from: https://data.worldbank.org/indicator/SL.UEM.TOTL.FE.ZS?locations=AE&view=chart.

UAE Embassy in Washington, DC (2018). Women in the UAE. Embassy of the United Arab Emirates, Washington D.C. Retrieved from: https://www.uae-embassy.org/about-uae/women-uae.

UAE Federal Competitiveness and Statistics Authority (2017). Labor Force. Federal Competitiveness and Statistics Authority, United Arab Emirates. Retrieved from: http://fcsa.gov.ae/en-us/Pages/Statistics/Statistics-by-Subject.aspx#/%3Fyear=&folder=Population%20and%20Social/Labor%20Force&subject=Population%20and%20Social.

UAE Ministry of Education (2018). Higher Education. UAE Ministry of Education, United Arab Emirates. Retrieved from: https://www.moe.gov.ae/En/OpenData/Pages/ReportsAndStatistics.aspx.

UAE National Elections Committee (2019). President issues resolution to raise women's representation in FNC to 50%. Retrieved from: https://www.uaenec.ae/en/news/details/40397.

UN Office of the Special Advisor on Gender Issues and Advancement of Women (2011). Gender Mainstreaming: An Overview. United Nations, Office of the Special Advisor on Gender Issues and Advancement of Women. Retrieved from: https://www.peacewomen.org/node/90256.

UNDP (2019). Table 5: Gender Inequality Index. United Nations Development Programme. Retrieved from: http://hdr.undp.org/en/composite/GII.

UNESCO (2017). Gender inclusion. United Nations Educational, Scientific, and Cultural Organization (UNESCO). Retrieved from: http://www.unesco.

org/new/en/social-and-human-sciences/themes/urban-development/migrants-inclusion-in-cities/good-practices/gender-inclusion/.

UNESCO (2018). Women in Science. United Nations Educational, Scientific, and Cultural Organization (UNESCO), FS/2018/SCI/51.

United Nations (2018). United Nations Secretary-General Appoints High-level Panel on Digital Cooperation. United Nations. Retrieved from: https://www.un.org/en/digital-cooperation-panel/.

Wajcman, Judy. (2010) Feminist theories of technology. Cambridge Journal of Economics, 34, 143–152.

West, Sarah Myers; Whittaker, Meredith; and Crawford, Kate. (2019). Discriminating Systems: Gender, Race, and the Power in AI. AI Now Institute, 1–33.

World Economic Forum (2017). The Future of Jobs and Skills in the Middle East and North Africa. World Economic Forum, 1–28.

World Economic Forum (2018a). The Future of Jobs Report 2018. World Economic Forum: 1–147.

World Economic Forum (2018b). The Global Gender Gap Report 2018. The World Economic Forum, 1–355.

World Economic Forum (2020). The Global Gender Gap Report 2020. The World Economic Forum, 1–371.

CHAPTER 11

The Art and Science of User Exploitation: AI in the UAE and Beyond

Helen Abadzi and Sahar ElAsad

1 Introduction

Thanks to free access, internet access has offered innumerable benefits. For example, an encyclopaedia of the world's knowledge is available at our fingertips. Tasks, such as banking transactions or library research that once took hours, are now doable in minutes. Users can also locate people of similar interests across the globe and find long-lost friends and relatives. But the road to these benefits has taken unexpected turns.

The inventor of the internet was Timothy John Berners-Lee, a British physicist and computer scientist. In 1989, while working at the European Organization for Nuclear Research in Berne, he proposed a project using hypertext to facilitate sharing and updating information among researchers. He built a prototype system in 1990 and, together with

H. Abadzi (✉)
University of Texas, Arlington, TX, USA

S. ElAsad
Regional Center for Educational Planning (RCEP), UNESCO, Dubai, UAE

E. Azar and A. N. Haddad (eds.), *Artificial Intelligence in the Gulf*,
https://doi.org/10.1007/978-981-16-0771-4_11

Robert Cailliau, created the first client and server implementations for what became the World Wide Web.

Initially, the internet aimed to deliver academic and military communications through desktop computers. Businesses and governments then became interested in the email and news services for their staff and financed its development. For commercial uses, the initial intent was to sell internet access on the basis of subscriptions, but the number of people in the world willing and able to pay has been limited. In the 1990s, companies tried to survive through fees and could not. Also the internet infrastructure over time proved extremely expensive and complex, far beyond what public entities could provide. Large-scale private investment was needed, along with much new research in software, hardware, and materials science. Around 1995, corporations developed a business model based on advertisements. Users could sign up for free email accounts in exchange for viewing ads; generally, users agree to such an exchange (Chatzithomas et al. 2014).

However, ad income proved insufficient for profits and expansion. As data science advanced around 2005, the business model evolved to obtain data from users in real time and to serve advertisements tailored to their needs. This meant that clicks on various links resulted in payments for the website hosts (Zimmermann & Emspak 2017). To maximize income, companies such as Google, Facebook, Microsoft, and Alibaba resorted to psychological research to develop procedures that keep users engaged online as long as possible (Wallace 2016). In the last ten years, corporations have conducted extensive experiments on users, varying the ratio and intervals of the exposure before reinforcement, to find out the conditions of maximal engagement.

Stories that are shocking or negative attract attention and are very useful. They may be false or exaggerated, but they receive prominence through software programming. The longer users are engaged, the more likely they are to become exposed to products and make purchases. Complex software programs have been commissioned to extract data from users' computers and smartphones, analyse them, and return relevant ads and more similar stories. The analyses include nudges to buy products that someone considered at a website but did not complete. Partnerships with communications companies have been developed for information storage and upload to users' computers and to company servers. Connectivity speed is crucial, so much advertisement has been addressed to users regarding their personal benefits of phones that use 3G, 4G, 5G.

The raw data and analytics are sold to paying clients. Since private companies own user datasets, profit becomes the primary criterion of their businesses, and their obligations are towards shareholders. Users have few rights to their data or realistic means to delete them. Tech companies and their affiliates make their own rules about data sales and change them as they see fit. They may promise users confidentiality but limit its scope in the small print, change policies without fully explaining their impacts to users, or simply make private preference settings hard to find.

The demand for personal data has fuelled a pattern of large-scale hacking of technical and personal information from companies across the world. Since 2016, news has frequently emerged about credit cards, bank accounts, or even airline data that were stolen from databases through sophisticated operations led by private companies or governments. The thefts have raised awareness of intermediary companies that hold people's personal information without their knowledge. Credit bureaus in the United States and elsewhere are for-profit companies that sell consumer data to advertising companies. The AI-based data thefts have resulted in publicizing citizens' private or unchangeable information, such as birth dates. This enables the development of elaborate schemes to deceive citizens.

The more the available variables, the more companies can maximize profits and minimize losses. For example, through identifying annual incomes, companies may price goods or services higher for some people who are considered ready or willing to pay more. The decisions are made by statistical methods that classify people in various categories. These programs are also able to learn and improve their predictive power. Decisions about individuals often have errors and biases, resulting in unfavourable decisions about hiring or renting apartments, but little can be done to deter them. As humans rely more on automatized decisions, they lose the power to override them.

The computational ability to store and link pieces of data resulted in alarming findings. When hundreds of variables become available for certain users, many features and preferences become predictable, and under certain conditions, users may change their views about products, social phenomena, or even politics. As such, persuasion became the cornerstone of many corporate efforts. In principle, any buyer with sufficient funds can hire companies that will buy users' data or access it, to later persuade users towards something or against it. People spending long hours on the internet are vulnerable to influence by news and products promoted by corporations. Persuasion tactics may include thousands of computer-generated accounts that are programmed to send human-sounding messages to give the impression of greater interest or agreement on a topic. Written words are not the only means of messages nowadays, AI algorithms can now produce believable pictures and videos such as Nvidia,[1] which cannot be reliably distinguished from fakes (Nightingale et al. 2017).

The ad-based model has created unforeseeable side effects in terms of human behaviours, economies, and democratic regimes. Personal information gives an outsized advantage to private parties and governments with agendas that range from marketing to regime change. Thus, the stock prices of technology companies have soared, and the value of data seems to exceed the value of oil today. Table 1 summarizes some of the benefits and costs of both users and technology companies.

The following section explores the neuroscience behind the persuasion tactics used by these businesses. The chapter will then discuss some of the strides and challenges of AI use globally and in the Gulf region and further explain the psychological effects of misusing AI. The chapter will finally conclude with some implications of AI across different disciplines.

2 Persuasion Strategies, Motivation Neuroscience, and Memory Functions

During World War II, propaganda specialists sought to identify how source, message, recipient, channel, and contextual factors affect a person's susceptibility. Various models were developed such as the Yale attitude change model (Hovland and Weiss 1951). This model focused attention to a message, its comprehension, and acceptance of a message. It subsequently became clear that there is a direct and an indirect route to information (the elaboration likelihood model by Petty and Cacioppo 1980). The direct route involves conscious arguments in favour or against various issues. This is useful to people who need the information, but this happens only in specific circumstances, usually when companies or governments want to persuade people who are not interested in the process.

[1] Learn more about Nvidia on their: https://www.nvidia.com/en-us/about-nvidia/.

Table 1 Benefits and costs for users and companies

Companies/ Users Interactions	*User Benefits*	*User Risks*	*Benefits to Companies*	*Risks to Companies*
Website content and attractiveness	Efficient replacement of letters and calls; community connectedness; timely information; entertainment; instant communication; tech jobs	Hiding communication purposes; discouraging privacy; narrowing users' information sources, spending inordinate time amounts	Profits; increased user willingness to study coding and hacking; availability of labour	Increasing calls for regulation and profit reduction; allegations of political interference and encouragement of violence
Products for sale	Users' stock and pension funds benefit from high stock prices	Psychological research used to turn users into big spenders	Companies sell stock and become "public"	Overvalued stock prices; user or worker complaints
Users as product	Users contact others of similar interests	False information, unfavourable comparisons create distress	User inputs constitute company products	Reputational risks; investigations; fines
Data harvesting	Users may look people up and learn private information, some of which is incorrect	Data used to calculate willingness to pay; influence users' social and political views; data hacking vulnerability to deceit	Value of personal data has exceeded the value of oil Potential for economic and political manipulation seems infinite	Much investment in "big data" infrastructure and analysis; industrial spying and conflicts; political conflicts

(continued)

Table 1 (continued)

Companies/ Users Interactions	*User Benefits*	*User Risks*	*Benefits to Companies*	*Risks to Companies*
Personalized pricing and targeting	No apparent benefit to users	Users pay more, have fewer choices, and may not know it	Companies merge or create alliances to minimize user options	No significant risks when competition is limited
Citizens' finances	Some websites offer financial advice; opportunities for distraction from financial or other problems	Ongoing income loss to corporate interests; inability to buy housing, repay debts, start a family; stress	Continuous revenue stream by inuring people to constant purchases	Some of the debt written off because of users' inability to pay
Political implications	Benefits to users from persuasion tactics are unclear, if any	Users may support political parties acting against their personal wellbeing	Feasible means to convert a portion of the population to clients' political views	Accusations of fake news, conspiracies; other parties must follow suit

This indirect route involves features of the communicator and the nature of the audience. Persuasive communication focuses on the credibility and attractiveness of the source, information early in a presentation, two-sided arguments for better-educated people and one-sided arguments for less well-educated people. People aged 18–25 seem more susceptible to persuasion (krosnick and Alwin 1989; Sears 1981). Research found that when we make a decision, we do not actually consider all the available information (Cialdini 2013). Instead, we use shortcuts based on social interactions such as reciprocity for an earlier favour, perceived scarcity of a product, authority of the person making arguments, consistency, how likeable the presenters are, and consensus among deciders. These methods are used extensively in marketing and social media and capitalize on community-based adaptations. Such shortcuts may have evolved because of working memory limits on the amount of information which can be processed rapidly. This indirect route to persuasion may create conditions for "mind control" or "social engineering". The indirect routes may take advantage of working memory overload, behaviour modification schedules, priming of concepts into implicit memory, or other routes (for a review see Abadzi 2020). These tactics give internet users no warning or opportunity to decide if they want to be convinced.

Cognitive dissonance may also help modify attitudes. A study conducted by Festinger and Carlsmith (1959), which involved small rewards or mild punishments, found that subjects were given a dollar to do boring tasks were more likely to report that they liked them than subjects given US$20 to lie. Because of cognitive dissonance, an internal justification arises, making subjects believe that they convinced themselves (Aronson et al. 2018). Beliefs are also changed through intentional priming. Recall accuracy of events showed bias according to the severity of the words used (Loftus et al. 1978), particularly of complex information. Priming people with words of various types and then showing them an event changes their views (Higgins et al. 1977; Aronson et al. 2018). For example, priming with hate words increases aggressive attitudes (Spanovic et al. 2010).

Distractibility is another persuasion technique that is used extensively on the internet. It is easier to persuade people who become distracted about an issue than people who have time to think about it (Albarracin and Wyer 2001; Festinger and Maccoby 1964; Aronson et al. 2018). People may make better decisions if they have time, as it has been proven that processing for only two extra minutes improves complex decisions

(Nordgren et al. 2011; Aronson et al. 2018). Cognitive overload and time pressure may cause users to believe false information and use it in making consequential decisions about a target. Interruption stops information processing and leaves people with false beliefs. This may be one reason why social media create constant disruptions, as constant notifications of new messages mean that multiple messages must be processed at the same time. Facebook Messenger (or WhatsApp, WeChat, or Snapchat for that matter) design their messaging system to interrupt recipients immediately by showing a chatbox, instead of helping users respect each other's attention. In fact, the captology models are built on these premises of "mass interpersonal persuasion".

Sales strategies often also rely on user distractibility. The corporations want us to click the purchase button quickly and think about the sale as little as possible. With credit cards stored and out of sight, purchases become painless. Corporations also turn purchases into games, in order to benefit from the thrill produced by the motivational system (see below). In principle in the United States, many pieces of merchandise can be returned later, but corporations may banish "serial returners" from getting refunds.

Another important persuasion technique involves action videogames. These give users little time to think, involve many dopaminergic actions, and are, therefore "addictive" (Alter 2018). Users learn the action game movements and skills easily, while self-control and critical thinking seem harder. In fact, the tendency of young people to spend hours daily into game worlds suggests the existence of evolutionary adaptations that are poorly understood. Arguably, the reinforcers have not changed from earlier generations, and people in many respects act as they did before the internet. Rewards and stimuli once came with snail mail and corded phones, and people did not sit on a stack of unopened letters. But messages are now delivered faster and more efficiently through the internet. It is easy and instant to click on the next stimulus with messages and other notifications. The real world cannot compete with the dopamine highs of videogames. Unless experiences are somehow modified, future generations may devote their greatest efforts to playing games, create their best memories in-game environments, and experience their biggest successes in-game worlds. (McGonigal 2011).

Motivational neuroscience offers some tentative explanations for the gamification thrills. The motivational system was originally designed to optimize energy use for animals that lived many millions of years ago, and it still weighs the costs versus benefits of each choice we make today. The mechanisms that bring this about lie in the limbic system, deep in the brain, where emotions reside. Areas, such as the striatum, amygdala, and mesolimbic pathway, are instrumental in processing and controlling hormones that create reward, drive, emotion, stress, and other facts that pour into motivation. Primary among those is dopamine, which signals an intent to explore further (Previc 2011). It is associated with excitement, whether positive or negative. Another is serotonin, which may be expressed in cases of defeat. It is high among dominant animals but also signals to keep calm and carry on. Epinephrin, endorphins, glutamate, and other chemicals also interact with the above.

The motivational system seems to have no absolute values; we can evaluate the desirability of something only if we see it in context (Kahneman 2011). For example, people who have lived in very poor circumstances become excited about foods and items that middle-class people may find trivial. One reason is the measure used, which is dopamine, as it shows differentiation between an expected reward and an actual reward. In that, our brains recognize that something important, whether good or bad, is about to happen, thus triggering the motivation to do something (Klein-Fluegge et al. 2016). These dopamine neuron activities also provide an index of reward prediction error (or gap), integrating of what might have happened with information on what actually happened. If an animal is pleasantly surprised, it is likely to continue carrying out the same action. If a person did not expect a positive outcome in a game yet received one, dopamine levels shoot up while serotonin goes down (Moran et al. 2018). Thus, the prediction gap between expected actions and reality is crucial. Nonetheless, the effect of neurotransmitters is momentary, as their effect wears off and the organism must repeat the action. In some respects, the motivational system of humans has been hijacked for entertainment. In real life, the system calculates costs and benefits, and people may receive a pleasant thrill if they manage to do something difficult. However, it is also possible to remove the costs in applications such as movies, roller coasters, and videogames. While the excitement is then kept and maximized, the outcomes of this are poorly understood in the long run.

Another component to the origin of internet behaviours is memory. It evolved to supply organisms with critical, just-in-time information needed for survival. It uses past information to predict the future (Klein et al. 2002; Todd et al. 2005). Attention and memory are extremely selective, designed to notice, store, and retrieve information that has the most importance for solving adaptive problems (Klein et al. 2002; Buss 2016). Memory storage and retrieval have costs, so animals have evolved functional specificity in memory (Garcia and Koelling 1966). Reasoning and memory depend on the specific circumstances. Thus we are more likely to remember information that helps decision-making in a specific location. Some types of information are more easily learned than others; for example, rats may associate food types with nausea but not noises (Garcia and Koelling 1966). People learn to fear snakes and spiders more easily than electrical plugs (Seligman and Hager 1972). Similarly, people seem prepared to learn some internet-related behaviours more than others. The ease with which certain internet-related behaviours are learned suggests complex applications of this ancient memory mechanism that are generally poorly understood. Humans inherited the memory features of earlier animals and integrated layers of complex reasoning. This means that a large segment of memory is unconscious. This includes both the working memory and the implicit memory functions. Therein lie many phenomena relate to internet manipulation. Unconscious, tacit, and implicit connections are made so that people react to but do not consciously process information. Moreover, the mind does seem to have a mechanism to keep track of unfinished business, which has been called the Zeigarnik Effect (Zeigarnik 1969; Savitsky et al. 1997). This may mean that people who did not finish a videogame may return to where they left off rather than start a new activity altogether.

Overall, there are links between distraction, time perception, and the phenomenon of getting lost for hours online. The need to respond fast and with little thought to the threats and opportunities of ancestral environments has resulted in certain cognitive biases. Lastly, rewards are linked to the release of dopamine in the limbic system, as the prediction gap is registered. The same way the environment trains organisms on associative learning tasks, such as learning to open doors, peel fruit, or avoid strong light. In this same manner, we also learn to associate the icons of various apps with specific events, such as incoming messages. Often, we also decide to stop other activities and click on these apps. Some internet psychologists call this training process “designing minds”, where we become designed to do as the companies bid.

3 Psychological Effects of AI Misuse

By designing minds to use directed content habitually, companies have opened a Pandora's box of human features. Some novel behaviours have arisen, that are hard to explain, including users prioritize attention to messages over work or less desirable tasks; they even respond to phone applications (apps) during social interactions and neglect real people; users are also always interested in presenting themselves well online, so they often exaggerate their looks or achievements. Elaborate deceptions seem common (Etgar and Amichai-Hamburger 2017); users, particularly women, may disclose much personal information to the world at large. Interlocutors often treat strangers as "friends", though they may never meet them in person. Serious mental health issues may arise when young women, in particular, feel slighted by their "friends"; users may find real-life less interesting than the excitement experienced on the internet. Given the relative values of dopamine prediction gaps, they may find real companions and lovers bothersome and trivial. Online sex may be more satisfactory. In fact, the deliberately created excitement of videogames may be creating delusions in some people.

And as mentioned above, users are losing control over their personal finance. Marketers take advantage of certain thinking biases, such as the tendency to view a product costing \$99.99 as cheaper than \$100. Physical and online stores abound with artificial discounts and merchandise presentations aimed at confusing buyers regarding the value or price of merchandise. These techniques are not new, but the opportunity of thinly veiled deceit is multiplied online. In addition to manipulating buyers' cognitive biases, marketing companies also try to improve sales through neuroimaging. They experiment with functional magnetic resonance imaging and eye-tracking that gives cues to the emotional states of prospective buyers. Thus products can be made irresistible. Thus the interests of various corporations are relentlessly pushed on consumers. Buyers are expected to act rationally and also earn enough to pay for all the merchandise pushed on them. They are responsible for mustering willpower to resist the techniques that very bright people set up to break down their willpower. There is practically no organized consumer protection. There are few activities aimed at shoring up consumers' financial literacy or helping them become disentangled from the above temptations.

4 Current AI Applications: The Good, the Bad, and the Ugly

In 2016, the United States (US) witnessed the election of a president with an extensive history in opinion manipulation. Coincidentally, the citizens of the United Kingdom (UK) were surprised to find out that the majority had voted to leave the European Union (EU). Despite the fact that in both countries social media users argued voraciously in different platforms against the US democratic candidate and against remaining in the EU, the news was still embellished by false or exaggerated news that circulated widely and appeared believable. This is due to the fact that arguments against the EU membership and against the democratic candidate were closely tailored to readers' beliefs and demographics. Demographically tailored campaigning has been used for decades in all countries. But some billionaires had hired Cambridge Analytica, a company that claimed ability to change opinions on the basis of large datasets "big data" that included personality variables. The company purchased millions of individual data records from Facebook and used them to convince people to vote for conservative candidates or to abstain from voting. The key was to find emotional triggers for each voter, for example, showing pictures of immigrants. These tactics were indeed successful (Liberini et al. 2018), as young people became suspicious of democratic processes and were more likely to vote for candidates who profess nationalism and disparage international cooperation.

In 2018, France was also repeatedly shaken by a movement of people against a planned petrol tax. Participants included men who destroyed property and battled police with weapons. Those groups had no apparent leader, but they were coordinated and organized through social media. Twitter accounts were strictly anonymous, and a number of them were identified as originating from right-wing organizations. Several messages attempted to intensify the anger and violence level stating falsely, for example, that protesters clamoured for the leadership of Donald Trump. Similarly, in Greece and in North Macedonia, 2018 was marked by extensive and angry movements regarding the name "Macedonia". They reflected efforts in both countries to prevent Former Yugoslav Republic of Macedonia (FYROM) from changing its name and joining the EU. Anonymous messengers on the internet hounded anyone who supported the agreement. News channels commentators mentioned vague threats from FYROM about irredentism and attacks on Greece, alarming the population. Even high school students took over schools and destroyed school property protesting the agreement about the name change. It was difficult to understand how the students had formed opinions about a complex historical and political problem. The force and tenor of objections against the North Macedonia name change made it unwise for anyone to express an opposing opinion, leading the national sentiment in both countries to be uniformly negative and aggressive.

Though few users actually kill, hostile attacks against parents, teachers, and friends have multiplied (UNESCO 2017). Aggression and disparaging remarks against women seem to have become more frequent in various countries. This outcome would be predicted from observational learning research dating from the 1960s (e.g. Bandura and Ross 1963), but the industry has succeeded in raising doubts about the effects and resists calls for change. In addition, under the guise of anonymity people may make disparaging comments and threats in social media. They may bully or hound users, attempting to drive some to suicide. They may send sexually explicit content and impersonate other people (Schradie 2019). Moreover, the increasingly aggressive tendencies may be contributing to the number of young men willing to engage in violent demonstrations. This raises the spectre of male groups desensitized by violent content since birth, fed distorted news with distorted frequencies from right-wing organizations, and fomenting uprisings against democratic regimes.

But not all AI uses are grim and evil; AI has the potential to provide insight into the learning process as well as into how to most effectively assess learning. It may also be utilized to strengthen the efficiency of teachers and change classrooms, or even "make them obsolete". Despite the promise of AI in education, it must be incorporated with care, as it could also hinder students' learning, for instance, through scaling the use of bad pedagogical practices (Ilkka 2018). AI in education would bring about transformational changes for teachers, as they become able to optimize the preparation of their educational materials. Machine language programs take content from teaching materials and translate them into chapter summaries, create practice tests, and study guides. Other systems help teachers and lecturers produce digital content for various devices, including videos and PowerPoint plug-ins. This includes translators that can produce subtitles in real time for what the lecturer/teacher is saying, thus allowing more students to benefit from top-notch lectures given in a language they do not master. Grading is another part of teaching, which teachers find stressful and time-consuming. For over a decade now, AI has already helped with the plagiarism checkers in detecting cheating, which have become more and more accurate through machine learning. In fact, there is now software which can detect authorship from a pool of writing styles database. Software that grade multiple-choice questions are now being supported by programs that can read and assess fill-in-the-blank questions. Moreover, new programs can now grade students' papers by "understanding" what types of questions are being asked, assess the student's answers, and even recognize handwriting. Lastly, Virtual Reality, which is one of the newest forms of AI, is affecting experimental activities by way of "augmented reality". These immersive experiences will allow students to perform trails and even go on field trips without leaving the classroom. As for tertiary educational institutions, the effect came most from software development, as needs have created a demand for coding expertise. Relevant skills software development turned out to be a relatively manageable skill to learn. Universities and technical institutes teach it, and coders mentor others on the internet. A worldwide push to teach young people coding reflects the current and future demand for coding skills, be it for legal or illegal software needs.

Regionally, and in October of 2017, the United Arab Emirates (UAE) Government announced the UAE Strategy for AI. The strategy is billed as the first of its kind and is projected to achieve the UAE Centennial 2071 objectives, enhance government performance through investing in AI adoption. The UAE's AI strategy covers development and application in nine sectors: Transport, health, technology, education, environment, space, renewable energy, water, and traffic. The UAE's Prime Minister and Ruler of Dubai, Sheikh Mohammed bin Rashid, said at its launch: "We want the UAE to become the world's most prepared country for artificial intelligence". As such, a couple of days after the launch of the AI strategy, the UAE became the first nation in the world to have a ministry and an associated minister dedicated entirely to AI. "The future is not going to be a black or white", said H.E. Omar Bin Sultan Al Olama, "as with every technology on Earth, it really depends on how we use it and how we implement it. People need to be part of the discussion". He added, "We will add clear laws, framework and roadmap for implementing AI to serve humanity, not control humanity".

AI in Education is one of the vital sectors in the government's new strategy; the goal is not only to reduce operational costs but also to increase the desire among students to learn. Some of the AI initiatives in the UAE Ministry of Education (MoE) involve an advanced data analytics platform totalling over 1.2 million students with more than 1,000 schools and more than 70 higher education institutions. The UAE MoE data analytics section is dedicated to developing e learning algorithms in support of strategic studies on the country's education system. This data analytics system entails scores from international assessments like the Program for International Student Assessment (PISA) and the Trends in International Mathematics and Science Study (TIMSS), in addition to data on performance reports, curricula, learning resources, financing, operations, teachers' professional development, teachers, students and even parents' feedback to produce fitting educational policies and strategies. The personalization of the learning experience is indeed the first great ongoing transformation of the educational system with the inclusion of AI. Several smart tutoring systems already exist, which track each student's ability in specific subjects and then tailor the content level and exercises to optimize the efficiency of learning. AI systems are similar but much more advanced in that its software continuously gauges the learner's comprehension and ability in complex subjects. While these intelligent systems are still in their early stages, within a few years, they will be powerful and will quickly spread through all educational systems worldwide. An AI system that has already managed to get its digital education platform into dozens of schools in Abu Dhabi and Alain is Alef Education. Although it was established only four years ago, Alef Education has worked closely with the UAE government bringing the platform to over 25,000 students in 57 public schools. It uses AI to tailor the curriculum to the individual demands of the students. The startup has made its first move into the United States, where its technology is used in two private schools in New York.

Nevertheless, AI applications have potential issues for the UAE and the region. On a personal basis, extreme gaming is very frequent among citizens and residents, and implications include neglect of families, distortion of personal lives, consumption of users' time, and influencing their opinions. Additionally, online engagement is creating new and more frequent behavioural phenomena. With screens taking primacy over daily tasks, people are also disclosing their most intimate details to the virtual world, or merely playing videogames for full days. Many are now considering people they have never met their close friends and get depressed due to aggressive or callous responses by these strangers. Young children, who spend much time on screens, are developing autistic-like behaviours, and school-age children are progressively exhibiting shorter attention spans.

These phenomena are consistent across countries and cultural variation seems rather limited. As a contributor to increasing childhood obesity in the UAE and Kingdom of Saudi Arabia, the rapid rise of mobile entertainment has also become a major public concern that involves AI in the region. A survey published by YouGov in 2018 found an association between the use of technology and childhood obesity, with 69% of the parents who were surveyed acknowledging this link, and two in five parents believing that their children would not choose a physical activity over a video a game. On the political realm, the implications are many and complex. Cyberwars are a concomitant feature to physical wars. Face recognition, military drones, attempts to make various installations self-destruct have by now well-established programming procedures. Given the political tensions in the region, the UAE has invested heavily in cybersecurity. In 2019, the Telecommunications Regulatory Authority (TRA) launched the UAE National Cybersecurity Strategy, which seeks to create a safer cyberinfrastructure in the UAE. The strategy will provide a regulatory framework for all cybercrimes in the country, protection to the country's critical assists, and a platform for global partnership in cybersecurity (TRA 2019).

5 Conclusion

When we talk about AI, we tended to think of robots doing smart advanced tasks on their own, mostly not now, but in the future. Fewer people think of Netflix movie suggestions, Google translation, Amazon book recommendations, scan-to-text software, digital personal assistants that find things online for us, manage our to-do lists, online orders, reminders, and all these other applications of AI. With AI becoming widely available recently, governments and institutions around the world are exploring its promise and pitfalls. Below are some of the positive and negative ethical, political, and health implications of AI.

5.1 Ethical Implications

The novelty of the technology can mask its persuasive intent. Persuasive technology can exploit the positive uses of computers. Computer software persists until a users' will breaks down. Computers control the interactive possibilities. Computers affect emotions but are invulnerable to them.

Computer software has no accountability. People cannot tell whether software impersonates a human (Fogg 2002). The business model and the very smart use of psychological research have opened Pandora's box, from which multiple demons have emerged. Worldwide millions of minds have been "designed" for habit-forming use and essentially become indentured servants to corporations that profit from their contributions. It is unclear how to retreat from this.

So what is to be done? There is first a need for an overarching framework of explanation that can link the multiple manifestations of the internet-based phenomena, including the desires of the corporate managers to exploit them. One promising framework is evolutionary psychology (see an overview in the Appendix). Yet, it is also possible to direct the videogame and some of the social media activity to alternative uses, to solve world problems (McGonigal 2011). Gamers are expert problem solvers and collaborators since they cooperate with other players to overcome daunting virtual challenges. There are games such as World Without Oil, a simulation designed to brainstorm, and therefore avert, the challenges of a worldwide oil shortage, and Evoke, a game commissioned by the World Bank Institute, which sent players on missions to address issues from poverty to climate change. Gamers could be able to leverage the collaborative and motivational power of games in their own lives, communities, and businesses. In reality, there is a dearth of institutions able and willing to direct their energies to uses other than military. Research, however, can be explained in educational interventions. For examples, students could be taught to be suspicious and to disbelieve information they read on the internet or to check facts. And since important behavioural rules are known, they could be taught to students explicitly. However, educational interventions require the engagement of conscious, explicit memory mechanisms. Given the constant information flow and the restrictions of working memory, many internet decisions are made through split-second, implicit memory processes, that are riddled with cognitive biases. It is essentially impossible to stop and think consciously about every item that is encountered. It is thus likely that large-scale educational interventions will educate people but still leave them vulnerable. It is unclear what other effective solutions exist.

Clearly, it is highly desirable to regulate data harvesting and opinion manipulation through countries' legal frameworks, which should have new privacy standards and proposed greater taxation of peddlers of online personal data. These frameworks should also regulate the use of AI algorithms with respect to decisions that affect the hiring, health, or welfare of individuals. But it is unclear which bodies can effectively regulate these new technologies.

5.2 *Political Implications*

The ability to persuade and change attitudes has been raised the bar for elections. Going forward, every politician must hire micro-targeters to influence potential voters; the same way that every musician must order fake YouTube views to become popular. The winners are conservative movements in this race, as they often have the billionaires with funds to hire programming companies for them. In recent years several authoritarian regimes have come to power, while democratic countries have gone under pressure from external forces to be put down. There is also a new vicious cycle that is emerging; young males, who naturally congregate in single-sex groups and practice cooperative competitiveness, are raised watching and playing violent media. They also have better than average chances of watching recruiters of various political and religious creeds on YouTube, given the tendency of human attention towards violent content. These may then become the proponents and soldiers of authoritarian governments to come. In addition, the sexualized content of the internet, linked to many videogames, may then be marginalizing women. If they are harassed sexually and physically, they may reduce their role or retreat from politics and daily life. They may thus leave the main roles to groups of males who will engage other groups in acts of war, as humanity has done for the millennia. Thus, all the gains of the mid-twentieth century in democracy, peace, and gender equality could sadly be reversed. Moreover, while people have lived under authoritarian regimes over the past centuries, they were also able to rebel against them and changed them. This may not remain possible for long. As technology improves, monitoring and control of individuals become easier. It is already possible to predict the likelihood of violent intent based on the mass of information collected through internet use. Technology is also improving the monitoring of thoughts, some of which is already possible. Further, if in 10–20 years, brain monitors were developed and implanted by authoritarian governments in the bodies of their citizens, the rebels planning regime changes may be apprehended even earlier.

5.3 *Health Implications*

A legion of ethical issues arises from people's data use and creation of compulsion to engage in social media. Many studies show mental and physical health effects that result from social media and videogame use. For example, frequent cell phone use linked to anxiety, lower grades, and reduced happiness in students (Lepp et al. 2014). In fact, heavy phone use is linked to oxidative stress (Hamzany et al. 2013). Many people play videogames safely, but there are problematic patterns online that are linked to social isolation and loneliness (Snodgrass et al. 2018).

Users often would like to disentangle themselves, but they need access to the internet for practical purposes, and turning access off is difficult. In many respects, the "big data" business model is responsible for the emergence of the internet phenomena. It is difficult to estimate the usage and "addiction" levels if engagement had remained relatively utilitarian, as it was until approximately 2005. The broad access to people worldwide through free accounts who might have never afforded private use has been inexorably linked to modifications of their behaviour that affect their lives in multiple ways. Excessive purchasing and undisclosed decisions about people on the basis of their data are costs that are often not considered.

It may be hard to believe but world populations have become more violent and less trusting of democracy through a model that originated in ad revenues. As AI becomes part of everyday lives, the internet that intended to bring universal knowledge to the world is unwittingly throwing us back into the Palaeolithic era. Now more than ever, humans ought to become more peaceful and content rather than be driven by ever-increasing emotion driven contests. Hopefully corporations and governments can control the undesirable effects in order to maintain sustainable consumption, world peace, and democratic regimes.

Appendix – Evolutionary Psychology: Essential Concepts

This branch of psychology examines mental functions from the perspective of human and animal survival. All organisms since the earth was formed survived and reproduced by adapting to the conditions around them. This is called natural selection. Those that could not leave viable offspring became extinct. Homo sapiens ultimately descends from single-cell organisms and a long line of intermediate animals. For survival, physical functions, as well as adapted behaviours, were necessary.

The behaviours of organisms evolved to adapt to the environments where they lived. Thus memory, motivation, social interactions, or personality features can be examined from the perspective of their utility for survival in humans and non-human animals (Tooby and Cosmides 1996; Buss 2016). The aim is to understand the processes that designed the human mind in order to resolve issues that are not easily explainable as independent phenomena (Piha 2018). Hypotheses informed by evolution can be tested (Durante and Griskevicius 2016).

Many behavioural adaptations probably developed in the Palaeolithic era, when humans were hunters and gatherers (about 2.5 million to 10,000 years ago; also called sometimes "stone age"). They experienced include nomadic, kin-based lifestyle in small groups, long life and low fertility for mammals, long female pregnancy and lactation, cooperative hunting and aggression, tool use, and the sexual division of labour. They also dealt with predators and prey, food acquisition and sharing, mate choice, child-rearing, interpersonal aggression, interpersonal assistance, diseases and a host of other fairly predictable challenges that constituted significant selection pressures. Successful behavioural adaptations arose over millennia of pressures, for example, to hunt large and dangerous animals.

The individuals carrying the most successful adaptations reproduced and left offspring that were more likely to display the traits that favoured survival. The advent of sexual reproduction facilitated the exchange of genes for survival in more challenging environments, so sexual selection has been a powerful selection force. The organisms able to mate with individuals possessing traits that favoured survival could have offspring that survived in subsequent generations. Genes carry DNA codes that influence various traits, so evolutionary psychology takes the "gene's eye view". If you were a gene what would you do to survive? In some respects, genes are agnostic. There seems to be no forward-thinking or general plan for better-adapted animals. The only criterion that matters is offspring survival to adulthood and reproduction.

Natural and sexual selection are not the only ways genes change. Chance variations over time (called genetic drift) also produce genetic expressions that may be harmful, harmless, or neutral. Some traits are merely "carried along" by an adaptive trait. Some are merely consequences of behaviours, such as the existence of belly buttons. Also, a trait can evolve because it served one particular function, but subsequently, it may come to serve another (exaptation). One example would be reading in humans, which uses circuits specialized for face recognition and the detection of small objects. It is often difficult to determine which behaviours are a direct outcome of evolutionary forces and which ones are carried along.

Survival strategies are manifest in all organisms. Even relatively simple organisms have the means to sense the environment and take action, avoiding harmful events and maximizing feeding. Nervous systems and hormonal molecules have generated mechanisms such as sensory and motor systems, memory and emotions, and motivation. And many animals found it advantageous to live in groups, such as bacterial colonies. The biological basis of behaviours is not always obvious, but existing research confirms many biological links.

Nevertheless, behaviours are not "hardwired" or genetically determined in a straightforward way. Multiple genes, as well as epigenetic mechanisms, typically affect behaviours. Humans are born with nervous systems that are malleable, and they become adapted to the environments they live in. And in various times of history, humans have encountered different cultures and technologies and adapted to those. Cultures also depend on human capabilities, so the possibilities are not infinite. Our behavioural systems today appear to function in ways that are quite different from those of 3000 years ago, but genes change slowly. People can only use capabilities whose essential features exist. For example, humans do not automatically use brainwaves to control equipment, though many people seem capable of learning this skill. And it is unlikely that the population can be trained to move large items from a distance, fly, or see behind walls.

In brief, the following are some evolutionary psychology concepts that seem relevant to internet phenomena.

Inclusive fitness is a concept which implies that people may support the offspring of relatives who carry their genes (Hamilton 1964). To avoid incest and expect support, humans (and many animals) privilege their closest relatives and give them much care (Trivers 1971). Evocations of brotherhood, kindred, lineage, or close communal links may increase care and support. One example is the many internet searches for long-lost relatives, genealogy, surnames, and genetic testing.

Unlike most animals, human males care for their offspring and defend them. To maximize their reproductive chances, they use complex long- and short-term mating strategies (Buss 2016). The typical female oestrus cycle of the animals was replaced by a hidden menstrual cycle, making it hard for men to detect fertility. Complex mating strategies ensued from this phenomenon, including a demand for fidelity, female coyness, and mutual attempts at deception.

One important strategy concerns the features of women who are to raise men's children. Men are primarily interested in young women, whose reproductive life is long (Artfolk 2017). The appearance of youth, therefore, has become a primary concern of women. Selfies attest to an interest in looking thin, to the point of digital manipulation. The desire for youth may be one reason for the emergence of child pornography.

Men must entice the most suitable women with resources suggesting the ability to support the young. They must also be considered worthy comrades. Important to male support strategy is the acquisition of resources and power that they then use to signal competence to women who need childbearing support and to attract male followers. One consequence is a tendency for males to be selfish and acquire power. Internet entrepreneurs and leaders have been almost exclusively male and have, among others, focused on power acquisition through users' data.

Humans have a mild degree of polygyny; women are about 18% smaller than men, and this dimorphism helps men prevail violently when opportunities arise. Women have had to adapt to abuse, while at the same time seeking strong males who will protect their families. Polygyny means that women must fight other women over men's resources. This puzzling phenomenon seems to be reflected in workplaces and in social media, where women tend to forgive straying men and attack other women.

Perhaps more than other mammals, humans have survived in communities. Groups were small, about 40–150 individuals, judging from contemporary hunter-gatherers. Arguably, people are unprepared to process the features of 1500 Facebook "friends".

Hunter-gatherer groups had a division of labour where women were occupied with child-rearing and gathering plants. Men hunted in male-only groups for animals that provided protein. Therefore, men have a bias towards single-gender groups.

Communal living in a precarious era seems to have given rise to multiple cognitive rules. To survive in groups of people who are not close relatives, humans follow a social contract among their members, delineating benefits and obligations on the basis of reciprocal altruism. People who coexist in communities are bound by social contracts, that is rules of cooperation for mutual benefit (Cosmides and Tooby 1992).

The collaboration between individuals tries following a tit-for-tat rule. Members must contribute as well as benefit. To ensure wise management of scarce resources, humans (and some primates) have developed abilities to spot and punish free-riders and cheats. In fact, a morality system ingrained in humans has developed for this purpose. The punishment or detection of deviants is regulated by social contracts. Revenge against violators causes pleasure. When social contracts do not exist, the punishment may be excessive (Cosmides and Tooby 1992).

Status is a powerful social force among animals and also humans. Skilful Palaeolithic hunters, stronger individuals, and those protected by more aggressive people left more offspring (Buss 2016). In principle, its role should be reduced in the modern era, but from a very early age, children seem attuned to status displays around them (Benenson 2014). Status seems to be established relatively quickly in schools and also in online groups. Men and women vie for status given their different roles, and marketers of various types take advantage of this desire. In an online environment, personal impressions are missing, but confident written expressions may serve to establish expertise, even when it does not exist (Locke and Anderson 2015).

Bullying seems to constitute efforts to increase status. It implies the use of superior strength and perception of greater benefits from a status increase. Higher status seems linked to credibility; that is, people may believe those who display indicators of high statuses, such as self-confidence and a number of allies. Women are often bullied online by both sexes, and much distress results from such events, particularly among adolescents.

Humans are also only one of two species that attack individuals of the same species. Thus, men have been embroiled in wars in just about every generation (Wrangham and Peterson 1996). Men seem specialized for fighting with their hands (Hare and Simmons 2019). One prominent feature of male attacks is the abduction of women, who will bear and raise the next generation. This feature has profound implications for sexual selection. Men fight with each other for the right to mate with women. Women vie and compete with other women for the attention of strong men who can protect and feed their children. One problem with this strategy is an increasing level of violence with subsequent generations, as women bear "sexy sons".

One consequence of male alliances towards hunting or abducting women is a rejection of women in their groups. They are less fit for battle, and they are certainly not interested in abducting more women. The male tendency for single-gender groups has been the exclusion of women from technological positions, as well as elaborate explanations about women being unfit for technology (as presented by a Google employee in 2017) (CBS News 2017). Sexual assaults and harassment may also be expressions of this ancestral feature.

Male fighting has had multiple effects in history. Men rely on others to follow orders, so they seem able to abide by strict hierarchies. Wars and conflict engender efforts to overcome the opponent, thus creating technological progress. Men compete among themselves and tend to be prejudiced towards men of other groups. But they also create alliances, a concept that is called "cooperative competitiveness" (Buss 2016; Benenson 2014). Men's ability to form alliances towards a greater goal is evident in the development of hardware and software, as well as in the advertisement alliances that collect and apply people's data.

Evidence for some of these traits comes from young children. Young boys show the signs of aggression and war-related play that parents may have never taught them (Benenson 2014). This may be one reason for the tendency towards motor expertise through videogames and the desire for multi-player videogames. Research also shows that strong men are less interested in equity (Price et al. 2017). Thus, groups of young males may espouse totalitarian political beliefs, and they may have increased aggressive displays as a result of violent media exposure.

The strategy for winning wars often leads to the use of deceptive tactics. Both genders use deception, but men are more likely than women to engage in various practices, and testosterone levels seem related to this propensity (Lee et al. 2015; Stanton 2017). The incidence of theft and deception that has prevailed on the internet that was to be open and egalitarian suggests trends that are not random. In effect, hacking is modern warfare. The strategies combine technical and psychological behaviours. They may hark to the role of testosterone, which increases the opportunity to trick people. The players on both sides find it stimulating, and two sides are needed for cooperative competitiveness. Evidence suggests that the male-only groups that created and promoted the internet treated its many challenges in terms of war-like offence and defence. Much of the terminology, such as "killer apps", attests to the male propensity for aggression.

Communal living and the decision mechanisms about it have big implications individuals inhabiting the internet. Like many animals, humans do not think alone about complex issues. Knowledge is distributed in a community, and it may be transmitted within a group through majority-biased learning (Sloman and Fernbach 2017). In fact, education is one strategy to disseminate community-based knowledge in an organized manner. One implication of communal knowledge is a tendency to believe what the majority says rather than doubt it. As mentioned above, our decisions are influenced by the number and status of people who have various opinions. Our brain collects statistics about these events. For people to understand a statement, they must have an initial belief about it. Understanding is believing (Gilbert et al. 1993). People may read or hear something, and then refute it through critical thinking. This feature creates a bias in favour of stated beliefs and gives an advantage to advertisers. Corporations create social illusions to manipulate the digital currency, which signals value to users. One means is fake accounts and images created to give the impression that many people support a certain position.

Social media heavily invest in the concepts of social contracts and reciprocal altruism (Trivers 1971; altruism consists of helping others at a cost to ourselves). Companies goad users to reply to messages and ensure that responses by default are open, enabling others to exert social control over this essential moral function. A user who refuses could be labelled a violator. But this requires a personal investment. By responding to comments, sending "liking" comments; one becomes a slave to the system and spends many hours creating content for advertising companies. People may feel good; certain brain areas are linked to social exchanges and linkage to the limbic system results in dopamine release (Purves et al. 2012). However, many interactions are generated by computers. Humans have had no relevant evolutionary pressures in distinguishing humans from software messages.

One important feature and community-based manipulation involves the "fear of missing out". People seem unwilling to be left alone, perhaps as an adaptation from millennia of community survival. Social media show friends engaged in interesting activities or purchases and websites emphasize that users must participate in order to keep up. Conformism is another important feature used, particularly when information is limited.

Women seem particularly prone to conformity (Pearson 1982; Nagle et al. 2014). And the current status of the internet, with written messages coming from unknown parties, seems to be a fruitful environment for this phenomenon. Marketing and news websites show the numbers of people who share or respond to certain messages, though the veracity of the numbers is uncertain. Social herding is expected, that is users are to follow the hints of others in opinions or in marketing.

Scarcity was a reality in the era before abundant food and basic infrastructure. Animals are set up to eat however much they can when they find food. The threat of scarcity, therefore, seems to be a powerful carry over into the modern era. Corporations market merchandise by evoking scarcity; "buy while supplies last", buy today. Evidence points towards a neural network that governs social interactions, and it's heavily linked to the mesolimbic dopamine pathway and therefore excitement and emotional centres (Supekar et al. 2018). Therefore, positive and negative interactions are rewarding. People find opportunities to improve status by humiliating others, such as taking compromising pictures and posting them online. These variables are not conscious, so people cannot easily say what they are doing and why. It may just seem "fun".

Humans learn a great deal from their companions. Children are able to imitate extensively, and there is a dedicated neurological circuitry for observational learning (e.g. mirror neurons; Rizzolatti et al. 2002). This suggests that they are likely to imitate high-status people similar to them. Aggressive and emotional expressions are sometimes beyond people's control, particularly when they have been trained from childhood by violent media.

One of the many applications for this pertains to "internet influencers", people who have large numbers of followers. Also, YouTube, owned by Google, is a prime medium for performance-related displays and seems to have become an important conservative recruitment strategy. As with all advertisers, viewers are encouraged to keep watching as long as possible, and people are drawn to content that is more extreme than their original position. So a business practice may be responsible for increasing the worldwide incidence of extreme views (Tufekci 2018).

The challenges among groups of males result in technological progress. Little by little over the millennia, humans have overcome the ancestral living conditions. We now live in warm housing, we are safe from predators, we have organized education, and have optimized food sources.

Thus, the inherited aspects of our mental apparatus interact with the environment in new ways that are only partly explored. Clearly, humans have not been subjected to aeons of natural selection for many features that are now common in our lives (Machin and Fisher 2015).

What helped survival and reproduction million years ago does not make for suitable behaviour in the twenty-first century (Harrari 2018; Kool and Agrawal 2016). People over the millennia interacted with others in person, rather than, say, as disembodied voices. They can detect deception in face-to-face interactions (ten Brinke et al. 2016), but not online. Also, the ancestral environment did not include pressures to distinguish between real and fake people. Accordingly, people cannot identify fake photos reliably (Nightingale et al. 2017). In person, people are often able to detect lying or duplicity, but they tend to believe statements seen in writing. Furthermore, people tend to conform when they are unsure of their role (Cialdini et al. 1990). Thus, they may become liable to suggestions. When a piece of software asks to fill out surveys or accept an invitation, it is difficult to conceive that there is no human behind the screen. Genetic linkages of traits in conjunction with new environments open a raft of possibilities that have not been encountered before. Clearly, the biological roots of behaviour are only partly understood, and new explanations are not always easy. Furthermore, humans display a lot of variability in their traits. But the uniformity of some behaviours like taking selfies (Diefenbach et al. 2017) shows the utility of exploring evolution in hopes of improving quality of life under these new circumstances.

References

Abadzi, H. 2020. Memory functions for excellence in the globalisation era. Cambridge Education Reform Reports, Cambridge University Press.

Albarracin, D. and Wyer, S. J. 2001. Belief formation, organization, and change: Cognitive and motivational influences. In D. Albarracin, B. T. Johnson, and Mark P. Zanna Eds.), Handbook of attitudes and attitude change, Mahwah, NJ: Erlbaum / 273–322.

Alter, A. 2018. Irresistible: The Rise of Addictive Technology and the Business of Keeping Us Hooked. Penguin Books.

Antfolk, J. 2017. Age Limits: Men's and Women's Youngest and Oldest Considered and Actual Sex Partners. Evolutionary Psychology January-March 2017: 1–9. https://doi.org/10.1177/1474704917690401.

Aronson, E., Wilson, T. D., Akert, R. M., et al. 2018. Social Psychology, 9th edition. Pearson Education.

Bandura, Albert; Ross, D.; Ross, S. 1963. Imitation of film-mediated aggressive models. Journal of Abnormal and Social Psychology. 66: 3–11. https://doi.org/10.1037/h0048687.

Benenson, J. 2014. Warriors and worriers. Oxford University Press.

Buss, D. 2016. Evolutionary Psychology. The New Science of the Mind. Routledge.

Cacioppo, J. T., and Petty, R. E. 1980), "Sex differences in influence-ability: Toward specifying the underlying processes," Personality and Social Psychology Bulletin, 6, 651–656.

CBS News (2017). Google worker says women don't advance in tech because of biology. https://www.cbsnews.com/news/google-worker-says-women-dont-advance-in-tech-because-of-biology/.

Cialdini, R. 2013. Influence: Science and Practice. 5th edition. New York, Pearson.

Cialdini, R. B., Reno, R. R., and Kallgren, C. A. 1990. A focus theory of normative conduct: Recycling the concept of norms to reduce littering in public places. *Journal of Personality and Social Psychology, 58,* 6), 1015–1026. http://dx.doi.org/10.1037/0022-3514.58.6.1015.

Chatzithomas, N., Boutsouki, C., Hatzithomas, L., and Zotos, G.. 2014. Social Media Advertising Platforms: A Cross-cultural Study. International Journal on Strategic Innovative Marketing. Vol. 1(2014) pages. https://doi.org/10.15556/ijsim.01.02.002.

Cosmides, L. and Tooby, J. 1992. Cognitive adaptations for social exchange. In J.H.Barkow, L.Cosmides and J.Tooby, *The Adapted Mind*, chapter 3, pp. 163–228.

Diefenbach, S. and Christoforakos, L. 2017. The Selfie Paradox: Nobody Seems to Like Them Yet Everyone Has Reasons to Take Them. An Exploration of Psychological Functions of Selfies in Self-Presentation, Frontiers in Psychology 2017. https://doi.org/10.3389/fpsyg.2017.00007.

Durante, K. M. and Griskevicius, V. 2016. Evolution and consumer behavior. Current Opinion in Psychology, 10, 27–32. https://doi.org/10.1016/j.copsyc.2015.10.025.

Etgar, S. and Amichai-Hamburger, Y. 2017. Not All Selfies Took Alike: Distinct Selfie Motivations Are Related to Different Personality Characteristics. Frontiers in Psychology, 26,(8), 842. https://doi.org/10.3389/fpsyg.2017.00842.

Festinger, L. and Carlsmith, J. M. 1959. Cognitive consequences of forced compliance. The Journal of Abnormal and Social Psychology, 58(2), 203. Retrieved fromhttp://web.mit.edu/curhan/www/docs/Articles/15341_Readings/Motivation/Festinger_Carlsmith_1959_Cognitive_consequences_of_forced_compliance.pdf.

Festinger, L. and Maccoby, N. 1964. On resistance to persuasive communications. Journal of Abnormal and Social Psychology, 58, 203–210.

Fogg, B. J. 2002. Persuasive Technology: Using Computers to Change What We Think and Do Interactive Technologies. Morgan-Kaufman Series in Interactive Technologies.

Garcia, J and Koelling, R. A. 1966. Relation of Cue to Consequence in Avoidance Learning. Psychonomic science 41):123–124. https://doi.org/10.3758/bf03342209.

Gilbert, D. T., Tafarodi, R.W., and Malone, P. S. 1993. You can't not believe everything you read. Journal of Personality and Social Psychology, volume 65, issue 2, pp. 221–233.

Hamilton, W. 1964. "The genetical evolution of social behaviour. I". Journal of Theoretical Biology. 7 1): 1–16. https://doi.org/10.1016/0022-519364)90038-4.

Hamzany, Y., Feinmesser, R., Shpitzer, T., et al. 2013. Is Human Saliva an Indicator of the Adverse Health Effects of Using Mobile Phones? *Antioxidants & Redox Signaling*, Vol 18, No. 6, p. 622. https://doi.org/10.1089/ars.2012.4751.

Hare, R. M. and Simmons, L. W. 2019. Sexual selection and its evolutionary consequences in female animals. Biological Reviews, 94, 929–956. https://doi.org/10.1111/brv.12484.

Harrari, Y. N. 2018. 21 Lessons for the 21st Century. New York: Spiegel & Grau.

Higgins S.T., Jones, C. and Rholes, W. 1977. Category accessibility and impression formation. Journal of Experimental Social Psychology 132:141–154. https://doi.org/10.1016/S0022-103177)80007-3

Hovland, C. I. and Weiss, W. 1951. The Influence of Source Credibility on Communication Effectiveness. Public Opinion Quarterly. Vo. 15, No. 4, pp. 635–650, https://doi.org/10.1086/266350.

Ilkka, T. 2018. *The Impact of Artificial Intelligence on Learning, Teaching, and Education*. European Union. Publications Office of the European Union.

Kahneman, D. 2011. Thinking Fast and Slow. New York: Farrar, Straus and Giroux.

Klein, S. B., Cosmides, L., Tooby, J. and Chance, S. 2002. Decisions and the Evolution of Memory: Multiple Systems, Multiple Functions. Psychological Review, Vol. 109, No. 2, 306–329.

Klein-Flugge, M. C., Kennerley, S. W., Friston, K., and Bestmann, S. 2016. Neural Signatures of Value Comparison in Human Cingulate Cortex during Decisions Requiring an Effort-Reward Trade-off. *Journal of Neuroscience*, 36, 39: 10002. https://doi.org/10.1523/jneurosci.0292-16.2016.

Kool, V. K. and Agrawal, R. 2016. *Psychology of Technology*. New York, Springer.

Krosnick, J. A.and Alwin, D. F. 1989. Aging and susceptibility to attitude change. Journal of Personality and Social Psychology, 57, 416–425.

Lee, J. J., Gino, F., Jin, E. S., Rice L., and Josephs, R. A. 2015. Hormones and ethics: Understanding the Biological basis of unethical conduct. *Journal of Experimental Psychology: General*, Vol. 144, No. 5, pp. 891–897. https://doi.org/10.1037/xge0000099.

Lepp, A., Barkley, J. E., and Karpinski, A. C. 2014. The relationship between cell phone use, academic performance, anxiety, and Satisfaction with Life in college students. Computers in Human Behavior, Vol. 31, p. 343. https://doi.org/10.1016/j.chb.2013.

Liberini, F., Redoano, M., Russo, A. et al. 2018. Politics in the Facebook Era Evidence from the 2016 US Presidential Elections. University of Warwick, Report no. No. 389.

Locke, C. C., and Anderson, C. 2015. The downside of looking like a leader: Power, nonverbal confidence, and participative decision-making. *Journal of Experimental Social Psychology*, 58, 42–47. https://doi.org/10.1016/j.jesp.2014.12.004.

Loftus, E., Miller, D.G. and Burns, H.J. 1978. Semantic integration of verbal information into a visual memory. *Journal of Experimental Psychology: Human Learning and Memory,* Vol. 4, pp. 19–31. https://doi.org/10.1037/0278-7393.4.1.19.

Lopez-Fernandez. O. 2015. Cross-Cultural Research on Internet Addiction: A Systematic Review. International Archives of Addiction Research and Medicine, Volume 1, Issue 2.

Machin A J, and Fisher D 2015. Why does engaging with families in maternal and infant care make such a positive difference? International Journal for Birth and Parenting Education 3: 19–23.

Machin, A. 2017. *Humans cannot deal with the internet phenomena because they never evolved to cope with them.*

McGonigal, J. 2011. Reality is Broken: Why Games Make us Better and How they Can Change the World. New York, NY, US: Penguin Press.

Moran, R., Kishida, K., Lohrenz, T. et al. 2018. The Protective Action Encoding of Serotonin Transients in the Human Brain, Neuropsychopharmacology, 43. https://doi.org/10.1038/npp.2017.304.

Nagle J. E., Brodsky, S. L. and Weeter, K. 2014. Gender, Smiling, and Witness Credibility in Actual Trials. Behavioral Sciences and the Law, 32(2), 195–206. https://doi.org/10.1002/bsl.2112.

Nightingale, S. J., Wade, K. A., and Watson, D. G. 2017. Can people identify original and manipulated photos of real-world scenes? Cognitive Research: Principles and Implications, Vol. 2, No. 30. https://doi.org/10.1186/s41235-017-0067-2.

Nordgren, L. F., Bos, M. W. and Dijksterhuis, A. 2011. The best of both worlds: Integrating conscious and unconscious thought best solves complex decisions. Journal of Experimental Social Psychology, 47(2), 509–511. https://doi.org/10.1016/j.jesp.2010.12.007.

Pearson, J. C. 1982. The Role of Gender in Source Credibility. Note: Paper presented at the Annual Meeting of the Speech Communication Association 68th, Louisville, KY, November 4–7, 1982. Retrieved from http://eric.ed.gov/?id=ED226390.

Piha, S. 2018. Evolutionary Psychology for Consumers: Awareness of ultimate explanations as a self-reflective tool for consumer empowerment. Economics Department, University of Turku.

Previc, F. 2011. The Dopaminergic Mind in Human Evolution and History. Cambridge.

Price, M. E., Sheehy-Skeffington, J., Sidnaius, J., and Pound, N. 2017. Is sociopolitical egalitarianism related to bodily and facial formidability in men? *Evolution and Human Behavior, 38*5), 626–634. http://dx.doi.org/10.1016/j.evolhumbehav.2017.04.001.

Purves, D., LaBar, K. S., Platt, M. L. Et al. 2012. Principles of Cognitive Neuroscience. Sinauer Associates.

Rizzolatti, G., L. Fadiga, L. Fogassi and V. Gallese. 2002. "From mirror neurons to imitation: Facts and speculations." In The imitative mind: Development, evolution, and brain bases, edited by A.N. Meltzoff and W. Prinz. Cambridge: Cambridge University Press, pp. 247–266.

Savitsky, K., V.H. Medvec, and T. Gilovich. 1997. "Remembering and Regretting: The Zeigarnik Effect and the Cognitive Availability of Regrettable Actions and Inactions." Personality and Social Psychology Bulletin 23: 248–257.

Schradie, J. 2019. The Revolution That Wasn't: How Digital Activism Favors Conservatives. Harvard University Press.

Sears, D. O. (1981). Life stage effects on attitude change, especially among the elderly. In S. B. Kiesler, J. N. Morgan, & V. K. Oppenheimer (Eds.), Aging: Social change (pp. 183–204). New York: Academic Press.

Seligman, M. E., and Hager, J. L. 1972. Biological boundaries of learning. East Norwalk, CT, US: Appleton-Century-Croft.

Sloman, S. and Fernbach, P. 2017. The Knowledge Illusion: Why We Never Think Alone. Riverhead Books.

Snodgras, J. G., Dengah, H., J., Lacy, M. G. et al. 2018. Social genomics of healthy and disordered internet gaming, American Journal of Human Biology. https://doi.org/10.1002/ajhb.23146.

Spanovic, M., Lickel, B. and Denson, T. F. 2010. Fear and anger as predictors of motivation for intergroup aggression: Evidence from Serbia and Republika

Srpska. *Group Processes and Intergroup Relations,* Vol 13, No. 6, pp. 725–739. https://doi.org/10.1177/1368430210374483.

Supekar, K., Kochalka, J. Schaer, M. et al. 2018. Deficits in mesolimbic reward pathway underlie social interaction impairments in children with autism. Brain. 2018 Sep 1;1419): 2795–2805. https://doi.org/10.1093/brain/awy191.

Stanton, Steven J. 2017. The role of testosterone and estrogen in consumer behavior and social and economic decision making: A review. Hormones and Behavior, 92: 155–163. https://doi.org/10.1016/j.yhbeh.2016.11.006. Epub 2016 Nov 11.

Ten Brinke, L., Vohs, K. D., Carney, D. R. 2016. Can Ordinary People Detect Deception After All? Trends in Cognitive Sciences, August 2016, Vol. 20, No. 8 http://dx.doi.org/10.1016/j.tics.2016.05.012.

Todd P. M., Hertwig R., and Hoffrage U. 2005. Evolutionary cognitive psychology. The handbook of evolutionary psychology pp. 776–802. Wiley, 111 River Street, Hoboken, NJ 07030-5774.

Tooby, J., and Cosmides, L. 1996. Friendship and the Banker's Paradox: Other pathways to the Evolution of Adaptations for Altruism. Proceedings of the British Academy. Retrieved August 26, 2018.

TRA. 2019. The UAE National Cyber Strategy. Retrieved from https://government.ae/en/about-the-uae/strategies-initiatives-and-awards/federal-governments-strategies-and-plans/national-cybersecurity-strategy-2019.

Trivers, R. L. 1971. The evolution of reciprocal altruism. The Quarterly Review of Biology, 461, 35–57. Retrieved August 25, 2018, from greatergood.berkeley.edu/images/uploads/Trivers-EvolutionReciprocalAltruism.pdf.

Tufekci, Z. 2018. YouTube, the Great Radicalizer. New York Times, March 10, 2018.

UNESCO. 2017. School Violence and Bullying: Global Status Report.

Wallace, P. 2016. The Psychology of the Internet (2nd ed). Cambridge University Press.

Wrangham, R. and Peterson, R. 1996. Demonic Males. Apes and the Origins of Human Violence. Boston: Houghton Mifflin.

YouGov. 2018. Childhood Obesity May be Linked to Excessive Use of Technology. Retrieved from http://www.mena.yougov.com/en/news/2018/07/16/childhood-obesity-may-be-linked-excessive-use-tech/.

Zimmermann, K. A., and Emspak, J. 2017, June 27. Internet History Timeline: ARPANET to the World Wide Web [Web log post]. Retrieved July 16, 2018, from https://www.livescience.com/20727-internet-history.html.

Zeigarnik, B.V. 1969. Introduction to Pathopsychology. Oxford, England: Moscow University.

CHAPTER 12

Fatwas from Islamweb.Net on Robotics and Artificial Intelligence

Julia Singer

1 Introduction

The discourse on the development and implementation of robotics and artificial intelligence (AI) is shaped worldwide by utopian ideas that transcend geographical and cultural borders. In the style of "traveling imaginaries" (Pfotenhauer et al. 2017), these utopian ideas are also vividly discussed in the Gulf countries, thus, creating hope for an economically prosperous future. In addition to these great hopes for AI's future achievements, there is also considerable interest in the field of robotics. It is striking that public, as well as scientific debate, often focuses on economic and technical issues in regard to these innovations. In doing so, it leaves the impression that AI and robotics are in general unrelated to religion, religious beliefs, and culture and that no link is required.

That this impression is misleading becomes clear when considering instances of technology intersecting religious beliefs. With regard to Islam and the use of artificial intelligence, this can be seen in at least four areas

J. Singer (✉)
Ludwig-Maximilians-University Munich, Munich, Germany
e-mail: julia.singer@lmu.de

E. Azar and A. N. Haddad (eds.), *Artificial Intelligence in the Gulf*,
https://doi.org/10.1007/978-981-16-0771-4_12

(hajj, Islamic banking, Quran, and fatwas), which I will later sketch briefly. Beyond the application of AI, there are also overlaps regarding Islam and robotics. Questions are raised as to whether robots (from an Islamic point of view) are considered human beings and thus are subject to Islamic law (Abdullah 2018). It is also critically discussed if the activity of robots in the food sector meets halal standards (Dahlan 2018) and after the robot Sophia received the Saudi citizenship 2017, it was (again) a point of controversy whether sex with a robot is to be interpreted as adultery from an Islamic legal perspective (Hānī 2019; Tijani 2012).

The application of AI and likewise of robotics in the religious field is progressing or at least is being considered. In addition to the utopian hopes associated with the use of these new technologies, questions arise that are closely related to human existence and/or to religious beliefs. Let us consider the potential impact on our own life if technology is not only physically stronger than us, but also outperforms us intellectually in many domains at the same time (strong AI). These questions about our worldview (*Weltanschauung*), the image of man and the image of God, and how they are changed by the construction and use of new technology are approached for some years from different perspectives (e.g. Collins 2018; Tegmark 2017).

The central question of this chapter is how Islamic religious scholars respond to AI and robotics. To the best of my knowledge, their ideas have never been investigated. A comparison of the three monotheistic religions and their perspectives on (new) technology would certainly be desirable but cannot be made in this chapter. While papers on Christian[1] and Buddhist responses to AI exist already (e.g. Varvaloucas 2018; Tamatea 2010; Promta et al. 2008), the Islamic perspective is almost unknown. By analysing fatwas, I seek to give a first insight into the way the topics AI and robotics are addressed by Islamic scholars.

In order to approach the Islamic debate, my chapter is divided into four main sections. I will begin by showing in which religious areas an application of AI and/or robotics is considered at all. There are also Muslim aspirations from think tanks (like "The Institute of Islamic Understanding

[1] The Christian discourse on AI is much more diverse and explicit than the Islamic one. Lukas Brand, Michael Burdett, and Noreen Herzfeld, among others, deal with it from a scientific perspective. Robert Geraci and Beth Singler work mainly in a comparative perspective and analyse how different cultures and religions react to new technologies like AI.

Malaysia") to treat the handling of AI from a religious perspective and initial publications are being prepared.[2] Similar to the German DIN or the international ISO, the focus is on standardizing the production and application of AI taking an Islamic perspective into account. In order to comprehend perceptions of robotics and artificial intelligence from an Islamic perspective, I will parse Islamic fatwas in the second section. Fatwas are legal opinions expressed by a Muslim scholar or anybody with expertise in Islamic Law (Arab. mufti). They give scholars the opportunity to react to innovations (social, legal, technological et cetera) from an Islamic point of view and to judge these according to Islamic law. In this function, muftis and fatwa councils are comparable with ethics commissions (i.e. institutions who give their assessment towards contemporary problems from an ethical perspective). I, therefore, use the analysis of these legal opinions as a first step to gain an insight into the scholarly discussion.

As a basis for the analysis, I refer in the third section to a corpus of 14 Arabic and English fatwas (Arab. pl.: fatāwā) issued between 2002 and 2019 on the web page Islamweb.net. They were all coming from the Qatari Ministry of Awqaf and Islamic Affairs, which is linked to the conservative Wahhabi branch of Islam. Since so far, no fatwas appeared on other websites, therefore a comparison of different Islamic positions is not yet possible. Separated by topic, fatwas dealing with robotics will be introduced in the first subsection, followed by fatwas focusing on artificial intelligence. I conduct the analysis by using qualitative content analysis and through the prism of the following three questions: To what extent can Islamic positions on AI and robotics be found in fatwas? What statements are made by the Islamic scholars? How does the treatment differ from robotics and AI?[3] The fourth and last section consists of a summary. Noticing that there is already a constant effort to implement new technologies in the religious realm, my thesis is that Islamic scholars are quite clear in terms of robotics but do not express a broad-based view on the

[2] The Institute of Islamic Understanding Malaysia (IKIM) is an Islamic Think Tank currently preparing a publication based on their round table discussion "The Ethics of Artificial Intelligence" (20.02.2019).

[3] At this point it is important to mention that I am not interested in whether something can be called AI in terms of the technical side. Rather, my focus is on how Islamic scholars talk about AI, how they react to it, and what views they take.

subject of AI. They tend to be rather reluctant to react but are open-minded about these technologies and not afraid of (intellectually) superior robots or artificial intelligence.

2 Use of Artificial Intelligence and Robotics in the Field of Islam

The aim of this chapter is to give an initial insight and an overview of how a conception of AI is already taking shape within the religious field. All the technologies dealt in this section are rather to be classified as weak intelligence. Nevertheless, they are examples of the fact that certain types of technology are already intersecting religious beliefs.

The Kingdom of Saudi Arabia, for example, attempts to better organize the hajj and umrah to Mecca. In order to prevent accidents caused by crowds, the organizers hope to be able to better analyse and predict the pilgrimage through the use of AI. The desire to make the hajj, which is an obligation in Islam, more effective through the use of new technologies is in line with the Saudi Vision 2030 intending to bring the country to a higher technical level.[4] There are likewise first attempts to improve the Islamic banking system utilizing AI. There were numerous initiatives of Islamic banks, especially at the end of the twentieth century, with the aim to offer their (mostly Muslim) customers products that would comply with the rules of the Sharia as it is forbidden to earn money through an interest in Islam (Ahmad and Hassan 2007). With regard to the Quran, there are also attempts to examine its content using AI in order to make it more comprehensible to the reader. In addition, some research focuses on using artificial intelligence to learn the correct recitation (*tajwīd*) of the holy text. The last area to be mentioned is the delivery of fatwas (Islamic legal opinions). While the Fatwa Center in the United Arab Emirates seeks to unify the published fatwas with the help of AI, a research group in Egypt is proposing a different approach. Due to the enormous number of inquiries regarding fatwas every day, they propose "an intelligent fatwa Questions Answering (QA) system that […] respond to a user's inquiry through providing semantically closest inquiries that were previously answered" (Elhalwany et al. 2015).

[4] See, for example, the official website for the "Vision 2030": https://vision2030.gov.sa.

As far as the practical use of robots is concerned, there are only a few areas of application. Apart from robots, which imitate the correct prayer postures (and particularly address children),[5] the debate about robots follows theoretical lines without directly influencing the practice or vice versa. Just as an example, Protestantism and the emerging theological debate about using a robot to donate blessings can be mentioned after the robot BlessU-2, constructed by the pastor Fabian Vogt, was used in the church congress in Wittenberg in 2017 (Grethlein 2019; Buse 2017). Also, in the Buddhist temple Kodaiji in Kyoto, a robot has been used since February 2019, which explains the Buddhist teachings to the believers. The humanoid robot named "Mindar" was commissioned by the temple itself and was developed together with Osaka University (Klein 2019).

3 Online Fatwas—Islamic Legal Opinions in Times of the Internet

Before I concentrate on the content of the fatwas I have analysed, a few brief comments on the fatwa genre and the website Islamweb.net are to be made. The status of the website and its Wahhabi ideological orientation is important to keep in mind, as it represents only a single Islamic position and is by no means a representative voice of Islam as a whole.

As mentioned before a fatwa is a legal opinion and can be issued by a Muslim scholar or anybody with expertise in Islamic law. While the process of giving such legal opinion is called iftāʾ, the person who asks for a fatwa is called mustafti and the person who answers mufti (Tyan and Walsh 2019). It is often a single person who turns to a mufti for clarification. The choice of the mufti means recognition of authority. This is emphasized by the view that a mufti is a representative of the Prophet Muhammad. Thus, the fatwa acquires a morally binding character for the questioner.

The basis for this (question and answer) system can already be found in several places in the Quran, in which the faithful ask the prophet Muhammad for information on various matters. After the death of the prophet and since the Sharia does not provide information and guidance on all aspects (of life), there was soon a demand for legal advice and

[5] Several examples of praying robots can be found on YouTube with the keywords "islam", "praying", and "robot".

the community of the believers resorted to the opinions of persons who were deemed competent (Tyan and Walsh 2019). Leaving aside the long history of the development of the genre of fatwa and of the iftāʾ process, it is noteworthy that media took upon a prominent role in the twentieth and twenty-first centuries in issuing fatwas. In order to distribute these fatwas, newspapers, radio, television, Internet and mobile phones are being used (Gräf 2019). While a fatwa was originally addressed solely to the person asking a question and related to their particular situation "they are now much more widespread [and] therefore, they are not only about individual cases anymore [...]" (Kutscher 2009). Websites are one of these new emerged medial formats, from which legal opinions can be requested and/or where you can find archives of fatwas. These sites can be administered by state institutions, non-governmental organizations, and even by independent scholars. From this multiplicity of sources follows a certain plurality of both authority and opinion (Gräf 2019). It remains difficult to assess what power and authority are to be ascribed to online fatwas. On the one hand (as in my case), the name of the mufti is often unmentioned. An evaluation of his authority and influence is therefore not possible. On the other hand, there is limited scientific research on the individual reception of this kind of fatwas.[6] It is usually unknown how mustaftis perceive their fatwas and whether or not they follow the recommendations given. Only the prominence of the website allows certain conclusions to be drawn about the impact of the issued fatwas.

An example of such an Internet-based website is Islamweb.net, the fatwas from which I will analyse below. The domain went online in 1998, making it one of the oldest websites publishing fatwas. Based in Doha, it is the website of the Qatari Ministry of Awqaf and Islamic Affairs and based on a rather Wahhabi-conservative interpretation of Islam (Gräf 2010). The legal opinions published there are thus related to the religious orientation of the Qatari state. The information regarding the fatwa committee varies depending on the language in which the website is displayed. While no specific names are mentioned on the Arabic-language page, the following information is given on the English page:

> In this site, there is a committee of specialists that is responsible for preparing, checking and approving the Fatwa. This committee comprises

[6] For the reception of fatwas (by Yusuf al-Qaradawi) and popularization of knowledge see: Gräf (2010).

> a group of licentiate graduates from the Islamic University, Al-Imaam Muhammad Bin Sa'oud Islamic University in Saudi Arabia, and graduates who studied Islamic sciences from scholars at Mosques and other Islamic educational institues (sic) in Yemen and Mauritania. This special committee is headed by Dr. 'Abdullaah Al-Faqeeh, specialist in Jurisprudence and Arabic language.[7]

Further information on the composition of the fatwa board could not be found. The page's up-to-dateness can be recognized by the fact that new fatwas are added on a daily basis as well by the fact that over the years it has been given a new design over and over again. The last major change of design was on the occasion of the 20th anniversary of the site in 2018 (Gulf Times 2019). In addition, information on page views makes clear that this is one of the websites that are regularly consulted and we can see that a lot of traffic is coming from Arab countries.[8] Since so far no fatwas on robotics and AI appeared on other websites, a comparison of different Islamic positions is not yet possible. It is likely that once the debate on this new technology intensifies, different Islamic perspectives will emerge.

4 Content Analysis of the Fatwas

The purpose of content analysis, in general, is to analyse material that originates from some form of communication. This chapter uses fatwas (as a form of communication) to understand what perspective Wahhabi scholars take on robotics and AI. The first step of the qualitative content analysis is to find out which opinions the scholars represent on artificial intelligence and robotics. In the second step, I question whether AI and robotics are treated differently, although the transitions between these technologies are fluid. The aim of the analysis is to gain insight into one part of the Islamic discourse through the medium of the fatwa. The special form of online fatwas is aimed at a wider audience through its accessibility and is also able to provide a platform for a debate on robotics and AI.

[7] See: https://www.Islamweb.net/en/fatawa/?tab=3.

[8] By using sites such as www.wolframalpha.com and www.similarweb.com, it is at least possible to assess the fatwa web pages in terms of visitor numbers and awareness, although, of course, all the presented figures are only estimates.

The sources for this survey are 14 different fatwas, of which ten have been written in Arabic and four in English. All of them were published on the web page Islamweb.net mentioned above. Surprisingly, this is the only website on which fatwas can be found on the subjects of artificial intelligence and robotics. My original intention to compare the different Islamic streams regarding these technologies was therefore not possible. Until May 2019, there were no fatwas on the two topics on the following fatwa-pages,[9] which contain the largest fatwa databases and represent different Islamic strands:

- Aboutislam.net (strong connection to Islamonline.net, fatwa database starts 2015/2016)
- Askimam.com (legal opinions issued by Ebrahim Desai, a South African scholar)
- Dar-alifta.org (publishes fatwas issued by state-run fatwa institutions, Egypt)
- Darulifta-deoband.com (founded 2007, India)
- Islamonline.net (founded 1997, former connection to Yūsuf al-Qaradāwī, Qatar)
- Islamqa.info (founded 1997 by Muhammad Al-Munajjid, a Saudi Salafi scholar)
- Sistani.org (founded by Alī al-Sīstānī, an Iranian Shia scholar)

I searched all these pages based on these English keywords (robot and artificial intelligence) and their Arabic equivalents (الربوت، الروبوت، الإنسان الآلي، الرجل الآلي، الذكاء الاصطناعي)— likewise without the article "al". The first fatwa on Islamweb.net is dated to the year 2002 and the last to 2019. More than half have been published over the past four years, which can be seen as an indication that both topics are becoming more relevant.

All of the Arabic fatwas seem to be edited by the fatwa board since almost no spelling mistakes are to be found in either the question or the answer provided. For that reason, we cannot be sure what the exact wording of the submitted questions was. Furthermore, they are all written in Modern Standard Arabic and do not contain any dialectical expressions,

[9] For some websites, the year of foundation remains unclear as well as the influential person(s) behind.

which also indicates that they were edited. The English ones, on the other hand, are less edited in terms of language and grammar.

Although the layout of the site has been modernized over the years, the outer structure of the fatwas themselves has not changed at all: they contain a short title (added later by a staff member or someone from the fatwa board[10]), an internal number, and the date of publication. Beneath and under the headline "the question" (*al-suʾāl*) the request of the mustafti is to be found and directly below under the headline "the answer" (*al-ijābah*) by the mufti is placed. Unfortunately, no information is provided about the mufti. No statement can, therefore, be made about his personal authority and influence. We know almost nothing about the mustafti likewise, except the submitted question contains personal information he has indicated himself. The response always begins with the fixed expression "Praise be to God, and blessings and peace upon the messenger of God, and his family and companions" and ends with "God knows best" (*allāhu aʿlam*), a traditional expression used as a gesture of modesty in the sense that, in the end, only God is to know the right answer. The following overview (Table 1) shows the total of 14 fatwas dealing with artificial intelligence and robotics.

The fatwas also vary significantly in terms of length and sophistication of questions and answers. While some questions are very concise and general, some deal with very specific aspects. There are questions that consist of a single sentence (see no. 1, 3, 13), while others ask for very specific information (e.g. no. 4 and 7) or are almost as long as the answer (see no. 5 and 8). With regard to the answer, there are also significant differences. While some fatwas indicate a variety of (religious) sources — citing passages of the Quran, hadiths, and statements of other scholars — on which their answer to the question relies (see e.g. no. 2, 7, 9), others are very brief (e.g. no. 3 and 13). As seen in the diagram below (Fig. 1) the Arabic fatwas reference to each other, whereas the English fatwas are not set in relation to each other. There is also no exchange across the two languages and none of these fatwas was translated from Arabic to English or vice versa.

The fact that AI is an issue that has received more attention only in recent years is also reflected in the fatwas: while the first fatwas that date

[10]This becomes clear as the title of the fatwa often receives element of the response of the mufti.

Table 1 Analysed fatwas*

No.	*Date*	*Title in English (partly translated)*	*Title in Arabic (if available)*
1	2002-07-26	(The automated human (the robot) … between prohibition and permissibility)	الإنسان الآلي (الربوت)...بين الحظر والإباحة
2	2003-05-11	Working with and using robots	-
3	2004-10-02	(Ruling on the use of robots in selling and buying)	حكم استخدام الإنسان الآلي في البيع والشراء
4	2007-11-28	(Ruling on placing robots on horsebacks during the race)	حكم وضع إنسان آلي على ظهور الخيل عند السباق
5	2009-11-15	(Ruling on drawing a robot)	حكم رسم الروبوت
6	2013-06-29	Using artificial intelligence	-
7	2016-05-14	Ruling on Transformer Toys	-
8	2016-07-25	(The difference in the creation of God and the creation of man)	الفارق بين خلق الله وصنع الإنسان
9	2016-12-14	(The enormous differences between the creation of God and the innovations of man)	الفروق الهائلة بين خلق الله وبين مبتكرات الإنسان
10	2017-03-09	Contradiction between Islam and reason impossible	-
11	2018-03-30	(The limits of human knowledge)	حدود العلم البشري
12	2018-12-23	(Artificial intelligence is not comparable to the creation of God)	الذكاء الصناعي لا يقارن بخلق الله
13	2019-02-01	(Is artificial intelligence forbidden?)	هل الذكاء الاصطناعي حرام؟
14	2019-02-17	(Is it possible to create a robot which thinks and possesses awareness?)	هل يمكن صنع رجل آلي يفكر ويمتلك إدراكًا؟

*The web links of the fatwas are provided in Appendix A

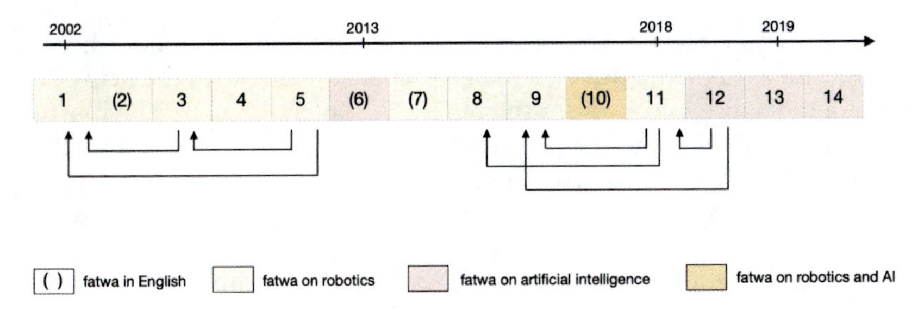

Fig. 1 Thematic overview of the analysed fatwas

to before 2016 focus exclusively on robotics, after 2016, it is rather AI that receives the most attention.

In order to examine the fatwas in terms of content, I will divide them into two categories according to the subject matter — robotics or artificial intelligence — while keeping in mind that these topics overlap. In the first step, I analyse the fatwas, which are related to the subject of robotics, and in the second step, those that deal with AI.

4.1 Fatwas on Robotics

Nine of the 14 fatwas I have found deal with robotics. Six of them have been written in Arabic and three in English, the first dating to 2002 and the last to 2018. While looking for concurrent fatwas, I was able to find them by using the English term "robot". As for the ones in Arabic, we can find several variants. Firstly, the English loanword *rūbūt* (روبوت) or *rubūt* (ربوت). About the same number of times the word "mechanical man" — *insān ālī* or *rajul ālī* (إنسان آلي أو رجل آلي)— is being used.

They cover an extremely wide range of topics and approach the subject of robotics in very different ways, but briefly summarized they address five aspects. The first aspect concerns the permissibility of making a robot in general and is closely related to the second aspect, namely how they should appear. That the issue of appearance is discussed at all finds its justification in the (alleged) ban of images in Islam.[11] The third aspect deals with the question of admissibility of the use of robots in certain

[11] Since this article does not provide the space for a detailed discussion of this topic, please refer to the bibliographies of the later mentioned encyclopedia articles or to the following further literature: Hawting (1999) and Noyes (2013).

activities. In the fatwas available to me, questions are asked about horse racing with a robot as a jockey, (foreign exchange) trade with the help of a robot and the use of robots for the purpose of entertainment. The fourth aspect concerns the question of the status of a robot with regard to the consideration that man is actually creating something as new as has only been done before by God himself. This act of creation begs the question of whether humans thereby put themselves on equal footing with God and to what extent their creation corresponds to the creation of God. The last aspect involves the question of whether, from an Islamic perspective, it is even possible to create a thinking robot. Or whether the knowledge (and the abilities) of man are limited in such a way that he/she will never be able to do so. On the following pages, the five aspects mentioned above will be examined more closely and will be shown on the basis of original passages from the fatwas.

More than half — namely five of the nine fatwas — contain statements regarding the question of whether it is permissible from an Islamic point of view to construct robots and how they should be designed. The first question — whether it is permissible to construct robots at all — is not answered literally by the Islamic scholars and does not occupy much space. But the tendency towards admissibility is clear for three reasons: first of all, no explicit ban on robots can be found in any of my fatwas, which strongly corresponds with the principle of the Sharia, that anything not explicitly forbidden is allowed. Secondly, the scholars emphasize the permissibility of using robots, as long as the intention of the user is lawful.[12] It is therefore unlikely that they would allow the use but prohibit the construction. Thirdly, there is a debate in the fatwas about what robots should look like. The discussion would certainly not be conducted by scholars if robots were not allowed to be produced at all from a legal perspective.

In one of the fatwas, the mustafti further raises the question of whether the construction of a thinking robot resembling a human is possible or whether the knowledge and therefore also the ability of humans are subject to everlasting limits.[13] The mufti is rather reserved and does not address this issue directly in his response, focusing instead on other aspects that have also been raised. At the end of his fatwa, however, he places a

[12] Fatwa no. 3 (2004).

[13] Fatwa no. 11 (2018).

literary quote from the Quran — the same one which was also mentioned by the mustafti in his question.[14] It indirectly and rather metaphorically contains the answer to the question of the possibility of creating a thinking robot:

> O men, a similitude is struck; so give you ear to it. Surely those upon whom you call, apart from God, shall never create a fly, though they banded together to do it; and if a fly should rob them of aught, they would never rescue it from him. Feeble indeed alike are the seeker and the sought! They measure not God with His true measure; surely God is All-strong, All-mighty.[15]

The core statement of the cited Quranic verse is that no one is able to create something akin to what God has done before. It, therefore, seems to be the case that the mufti does not believe that the creation of a thinking robot would be possible. He himself does not give an interpretation or further explanation. The meaning of the verse is therefore open for the interpretation by the mustafti.

On the question of the design of a robot, the fatwa committee comments more explicitly and even if they are not asked about it directly. The fact that this topic receives a lot of attention lies in the idea of the "prohibition of images" in Islam. Very briefly summarized, there is no consensus among the Islamic schools of law regarding this issue, but Wahhabism represents a rather strict interpretation of the religious sources. The "ban of images" derives from the Quran and is based on the linguistic correspondence between the two verbs "to create" (*baraʾa*) and "to fashion/form" (*ṣawwara*). The actions represented by these two verbs are each attributed to God, and God thereby becomes the creator (*al-bāriʾ*) and the fashioner (*al-muṣawwir*) at the same time. In conjunction with several passages from the hadiths,[16] it follows "that all human fashioners are imitators of God and as such deserving of punishment". According to a certain interpretation of the religious sources, a kind of norm of prohibition to copy living beings who have a soul (*rūḥ*) has emerged (Wensinck and Fahd 2019). Through passages from the hadiths,

[14] Surah 22 (al-ḥajj): 73.

[15] Surah 22 (al-ḥajj): 73–74. Translated by Arthur Arberry (Arberry 1955).

[16] For the exact passages in the quran and the hadiths, please refer to Wensinck and Fahd (2019).

the handling of such images is regulated and unlike often assumed there is no call for a complete deconstruction but jurists advise instead “a partial erasure of the depicted figure” (Flood 2019).

With regard to the question of the design of robots, the Islamic scholars warn against a too human representation. Most of the mustaftis are also aware of this issue and explain in their question — without being forced to — very detailed the design of the robot in detail and/or specifically ask, whether the “prohibition of images” regarding the appearance has to be considered. Although the fatwas are sometimes far apart in terms of time, a consensus can be seen in the scholars’ answers: they all state that robots constructed similar to living beings, who have a soul, are not permissible. They consider the attempt of such a construction either as a bravado of the skills of the constructor[17] or warn against imitating God’s act of creation.[18] However, they mention as an exception that they are permissible if the representation is not complete (*illā idha kāna taṣmīmuhu ghayra muktamil*).[19] In order to reach this state, three of the fatwas suggest that the head of the robot should be removed.[20] The removal of the head is in accordance with the partial erasure proposed by the religious scholars mentioned above.

There is a clear difference between the three Arabic and the two English fatwas, which deal with the question of design. While in the Arabic fatwas no references to sources (such as hadiths or statements by scholars) are made, such references are common in the English fatwas. Both in “working with and using robots” and in “ruling on transformer toys” in each case three hadiths are mentioned and in the later fatwa the statements of the scholars Muḥammad ibn al-ʿUthaymīn (1925–2001) and Ibn Qudāmah (1147–1223) are quoted. It seems as if an Arab reader would need less of such source references than an English one. However, it should be noted that these references are only indicating the expertise of the mufti because they are difficult to verify. All of the provided sources are treated like the following passage:

[17] Fatwa no. 1 (2002).

[18] Fatwa no. 2 (2003) and no. 7 (2016).

[19] Fatwa no. 1 (2002).

[20] Fatwa no. 2 (2003), no. 5 (2009) and no. 7 (2016).

> The Prophet (Sallallahu Alaihi wa Sallam) said: "The most grievous torment for the people on the Day of Resurrection would be for those who try to imitate Allah in the act of Creation." [Reported by Imam al-Bukhari].[21]

Without further (bibliographical) reference, the mustafti must trust that somewhere in al-Bukhari (810–870) this saying of the prophet can be found — but he cannot consult the passage himself.

The use of robots is discussed in seven fatwas, which all treat very specific topics. They concern horse racing with a robot as a jockey (foreign exchange), trade with the help of a robot, the use of robots as transformer toys and drawing them for the purpose of entertainment. Without going into the details of these very specific questions, it can be observed that the use of robots, in general, seems to be unproblematic. As long as the rules regarding the design are observed and as long as the purpose behind the usage is permitted (*mubāḥ*),[22] the scholars have no objection.

The question of whether the creation of a robot is to be related to God's act of creation is closely linked to the question of design mentioned above. The scholars urgently warn against creating something that resembles a human or a living being, since this imitation of something already existing comes close to the act of God's creation.[23] Their recommendation to Muslim designers is that they should stay away from such constructions for safety's sake, in order to commit neither a "sin" nor a "wickedness" (*ḥattā lā yadkhulu fī al-ḥarj wa-al-ithm*).[24]

A slightly contradictory argumentation can be found in a fatwa from 2016 titled "The enormous differences between the creation of God and the creation of man" (*al-furūq al-hāʾilah bayna khalq allāh wa-bayna mubtakarāt al-insān*). The legal information requested here addresses the question of inventing an artificial organism (which possesses robotic and animal parts) challenges the uniqueness of God, namely his power of creation. In this fatwa, the mufti does not advise that such research should be better avoided. He denies instead in general that human inventions could come close to God's creation in any way. According to him,

[21] Fatwa no. 2 (2003).

[22] The Arabic term "mubāḥ" (مباح) is the third category of the five decisions (al-aḥkām al-khamsah) by which actions are judged.

[23] Fatwa no. 1 (2002), no. 2 (2003) and no. 7 (2016).

[24] Fatwa no. 1 (2002).

all human inventions would be merely attempting to imitate something already existing in nature. Although he adds that nature itself was created by God, there is no indication that humans should refrain from engaging in the process of creation.

Roughly summarized, there are two strategies for dealing with the question of whether robotics comes close to God's creation. The first strategy is that the Muslim constructor should be more careful in the choice of design. With replicas of living beings (with soul) there could be an approximation to God's act of creation. The second strategy tends rather to the consideration that an imitation of God is basically not possible — the question of design is then no longer mentioned at all.

Another fatwa, which deals with a quite similar question as the two just mentioned, is focusing on whether a robot could be in his construction better than God's creation. The question here is not whether man can approach God by building robots, but whether he is even able to surpass him regarding the act of creation. The mufti avoids the question by merely talking about what it means when machines are smarter than humans:

> If we presume that some machines are more intelligent than man, then this does not undermine religion at all. Rather, these machines are man-made, and the fact that man was enabled to manufacture them is a favor from Allah upon him. So how can something that requires being grateful to Allah lead to disbelieving in Him?[25]

If people construct something that surpasses them in capabilities, this is not seen as a threat to the creation of God.

4.2 *Fatwas on Artificial Intelligence*

The remaining fatwas, four of them, deal with the topic of artificial intelligence. Three of them have been written in Arabic and have been published in the last two years. Based on the information in the question, it becomes clear that two of the three fatwas were requested by the same mustafti asking for more detailed information regarding his original question. The one written in English can be dated back to the year 2013.

[25] Fatwa no. 10 (2017).

The term "artificial intelligence" *dhakāʾ iṣṭināʿī* (ذكاء اصطناعي) is a literal translation of the English term.

In contrast to the fatwas on robotics, they do not cover a variety of topics and the questions are not very specific either. The mustaftis either want to know if (the use of) artificial intelligence is forbidden in Islam, without giving the mufti any further detail or they ask if the simulation of the human brain and thereby the construction of a thinking machine is possible.

Briefly summarized, we can identify three issues that run through all of them: the first aspect is the question of permissibility which is (again) closely linked to the principle of the Sharia, that anything not explicitly forbidden is allowed. The second aspect is the definition of artificial intelligence itself and it becomes quite clear that the scholars either tend to ignore the implications of a strong AI or do not believe in its possibility. The third aspect is not very elaborated but addresses the connections between God/God's creation, humans and the development of artificial intelligence.

The admissibility of AI depends both on the principles of the Sharia and on the use. The treatment of AI as a new technology is therefore very similar to the treatment of robotics. Regarding the Sharia the scholars state:

> Amongst the important basic rules and principles of the Islamic religion is that all things are permissible except what involves an Islamic prohibition. [...] Islam came to make humans happy and prescribed everything that benefits them. Therefore, there is nothing wrong with using such technologies that you have called artificial intelligence to make machines help in fulfilling some of the interests of people.[26]

Technology as a tool for fulfilling human interests is thus basically permitted. The concept that Sharia basically allows everything that is not explicitly forbidden is not only applied to robotics but also to AI. In the Arabic fatwa this approach is also represented and basically transferred to all new technologies:

> The Muslim, if he had understood this rule, is freed from the obligation to ask about each innovation in particular on which the people know is not

[26] Fatwa no. 6 (2013).

> forbidden in itself [...] like in general technical matters including artificial intelligence.[27] [translation by the Author]

As we can see above, the admissibility granted by the Sharia is regulated in a further step by the later use. If the use of this device is permitted under Islamic law, it is also the (new) technology.[28]

The passages quoted above illustrate a certain openness of the scholars towards the new technology, but in the remaining text, it becomes evident that they don't share their understanding or their conception of AI. The scholars do not take up the current distinction between a weak and a strong AI and reject instead the possibility that a machine can go beyond its capabilities programmed in the first place:

> In the creation of machines with this kind of intelligence, the machine does not go beyond its machine existence and it has no properties of living beings at its disposal.[29] [translation by the Author]

From the scholars' point of view, a machine can only do what it was previously explicitly taught by programming.[30] From this non-differentiation between strong and weak AI, it follows that they do not see people endangered by machines:

> Even though it (the field of research) become stronger and develop a lot, [artificial intelligence] will not go beyond what a machine does or attack human abilities.[31] [translation by the Author]

This statement is strictly speaking a contradiction to the statement in Fatwa no. 10, which has already been discussed in the section "Fatwas on Robotics". The scholars assume in that particular fatwa that machines are in some respects more intelligent than humans, but do not consider this as problematic.

Finally, compared to the statements on robotics, scholars are less concerned that through the development of artificial intelligence, humans

[27] Fatwa no. 13 (2019).

[28] Fatwa no. 6 (2013) and no. 13 (2019).

[29] Fatwa no. 12 (2018).

[30] Fatwa no. 14 (2019).

[31] Fatwa no. 14 (2019).

challenge God's power of creation. On the contrary, they strongly reject that the development of AI could threaten the uniqueness of God:

> The field of this study (Artificial Intelligence) is in no way in conflict with the uniqueness of God in regard to his creation. Nothing in it corresponds to the qualities of the almighty Creator.[32] [translation by the Author]

Unlike the robot, which can become human through his appearance and resemble God's creations, the scholars do not consider AI in any way as alive. In particular, it is the existence of the soul (*wujūd al-rūḥ*) coming directly from God, or in the case of the machine, the lack thereof why the scholars refuse to consider AI as a living being. However, it is striking that scholars tacitly ignore that the possibility of a strong AI aware of itself and independent of its programming is or could at least theoretically be possible. At this point, the issued fatwas clearly differ from the concerns or uncertainties that become evident in the questions of mustaftis. Their questions were clearly aimed at the emergence of a strong AI and whether it is becoming a threat to humanity, and to what extent its abilities come close to (or even surpass) those that humankind has received from God.

5 Conclusion

The analysis of Islamic positions on artificial intelligence and robotics in fatwas has shown that although the discourse on these technologies emerged already in 2002, it was not conducted in much depth in this genre. I was only able to find 14 fatwas with a strong preponderance that deal with the subject of robotics and only a few, especially in recent years, are dealing with AI. They were all published on a single website (www.Islamweb.net), which is connected to the Qatari Ministry of Awqaf and Islamic and therefore belong to the conservative Wahhabi branch of Islam. Therefore, a comparison between the different Islamic currents is not (yet) possible.

Nevertheless, based on the content analysis, three statements about the attitude of the Wahhabi oriented scholars can be made: (1) with regard to the treatment of the two technologies, the discourse on robots seems to be much more versed in Islamic law. Only in the fatwas about robotics can numerous hadiths, passages from the Quran, and statements of other

[32] Fatwa no. 12 (2018).

scholars be found. By treating the question of design, the muftis build their own argumentation on the sources just mentioned. Since there are overlaps with robotics and Islamic law, scholars have a fairly explicit and clear stance on this technology, because they can, in order to answer the question of the mustaftis, refer to different religious sources and build argumentation based on them. With regard to AI, it is noticeable that Islamic scholars do not refer to any external sources. They limit their statements mostly to the principle of the Sharia that everything is allowed unless it is explicitly forbidden. (2) On the basis of the questions raised, a slight uncertainty on the part of the mustaftis becomes apparent: they fear that the construction of robots or AI imitates God's creation and therefore explicitly ask about the admissibility of the new technology based on the Sharia. The muftis, on the other hand, do not share this concern (except for the design of robots). They see neither the creation nor the use of these technologies as a threat to man or as a threat to the uniqueness of God. (3) However, this open-minded attitude is closely related to the fact that they tend to avoid the really tricky questions and leave them unanswered. In this way, they do not discuss the question of whether robots will be a better version of God's creation or what strong AI would mean for humanity. The scholars neither reference to religious sources nor do they use fatwas to disseminate their deeper understanding regarding the treatment of artificial intelligence. They remain rather reluctant to issue a detailed opinion about artificial intelligence as a new technology yet.

Acknowledgements This chapter is part of my doctoral research which focuses on "Media Representations of Robotics & Artificial Intelligence in the Gulf Region" under the supervision of Prof. Andreas Kaplony. I am also very grateful to Martin Šotola for the proofreading and his helpful feedback.

Appendix A – List of Fatwas with Web Links

- Fatwa no. 1 - الإنسان الآلي (الربوت)...بين الحظر والإباحة - issued 26.07.2002. online available: http://www.Islamweb.net/ar/fatwa/20017/.
- Fatwa no. 2 - Working with and Using Robots - issued 11.05.2003. online available: http://www.Islamweb.net/en/fatwa/85827/working-with-and-using-robots.
- Fatwa no. 3 - حكم استخدام الإنسان الآلي في البيع والشراء - issued 02.10.2004. online available: http://www.Islamweb.net/ar/fatwa/54124/.

- Fatwa no. 4 - حكم وضع إنسان آلي على ظهور الخيل عند السباق - issued 28.11.2007. online available: http://www.Islamweb.net/ar/fatwa/102051/.
- Fatwa no. 5 - حكم رسم الروبوت - issued 15.11.2009. online available: http://www.Islamweb.net/ar/fatwa/129139/.
- Fatwa no. 6 - Using Artificial Intelligence - issued 29.06.2013. online available: http://www.Islamweb.net/en/fatwa/211585/using-artificial-intelligence.
- Fatwa no. 7 - Ruling on Transformer Toys - issued 14.05.2016. online available: http://www.Islamweb.net/en/fatwa/323455/ruling-on-transformer-toys.
- Fatwa no. 8 - الفارق بين خلق الله وصنع الإنسان - issued 25.07.2016. online available: http://fatwa.Islamweb.net/ar/fatwa/331972/.
- Fatwa no. 9 - الفروق الهائلة بين خلق الله وبين مبتكرات الإنسان - issued 14.12.2016. online available: http://www.Islamweb.net/ar/fatwa/341734/.
- Fatwa no. 10 - Contradiction Between Islam and Reason Impossible - issued 09.03.2017. online available: http://www.Islamweb.net/en/fatwa/347275/contradiction-between-islam-and-reason-impossible.
- Fatwa no. 11 - حدود العلم البشري - issued 30.03.2017. online available: http://www.Islamweb.net/ar/fatwa/349579/.
- Fatwa no. 12 - الذكاء الصناعي لا يقارن بخلق الله - issued 23.12.2018. online available: http://www.Islamweb.net/ar/fatwa/388999/.
- Fatwa no. 13 - هل الذكاء الاصطناعي حرام؟ - issued 02.01.2019. online available: http://www.Islamweb.net/ar/fatwa/389512/.
- Fatwa no. 14 - هل يمكن صنع رجل آلي يفكر ويمتلك إدراكًا؟ - issued 17.02.2019. online available: http://www.Islamweb.net/ar/fatwa/392146/.

References

Abdullah, Shahino Mah. Intelligent Robots and the Question of their Legal Rights: An Islamic Perspective. Islam and Civilisational Renewal 9, no. 3 (2018): 394–397.

Ahmad, Abu Umar Faruq and M. Kabir Hassan. Riba and Islamic Banking. Journal of Islamic Economics, Banking and Finance 3, no. 1 (2007): 1–33.

Arberry, Arthur. The Koran Interpreted. 2 vols. London: Allen & Unwin, 1955.

Buse, Uwe. Herr Roboter, Gib Uns Deinen Segen. Spiegel Online. (2017-09-04). Consulted online on 02 May 2019. https://www.spiegel.de/spiegel/

kirche-in-wittenberg-menschen-lassen-sich-von-einem-roboter-segnen-a-1162977.html.

Collins, Harry. Artifictional Intelligence: Against Humanity's Surrender to Computers. Medford, MA: Polity Press, 2018.

Dahlan, Akbar Hadi. Future Interaction between Man and Robots from Islamic Perspective. International Journal of Islamic Thought 13 (2018): 44–51.

Elhalwany, Islam, Ammar Mohammed, Khaled Wassif and Hesham Hefny. Using Textual Case-Based Reasoning in Intelligent Fatawa QA System. The International Arab Journal of Information Technology 12, no. 5 (2015): 503–509.

Flood, Finbarr Barry. Iconoclasm. In Encyclopaedia of Islam, THREE, edited by Kate Fleet, Gudrun Krämer, Denis Matringe, John Nawas and Everett Rowson. Consulted online on 02 May 2019. http://dx-1doi-1org-1ffotf5190719.emedia1.bsb-muenchen.de/10.1163/1573-3912_ei3_COM_32363.

Grethlein, Christian. Sollen Roboter Segnen? mdr (mitteldeutscher rundfunk). (2019-01-24). Consulted online on 02 May 2019. https://www.mdr.de/religion/kirche-digitalisierung100.html.

Gräf, Bettina. Fatwā, Modern Media. In Encyclopaedia of Islam, THREE, edited by Kate Fleet, Gudrun Krämer, Denis Matringe, John Nawas and Everett Rowson. Consulted online on 02 May 2019. http://dx-1doi-1org-1ffotf5lm01f8.emedia1.bsb-muenchen.de/10.1163/1573-3912_ei3_COM_27050.

Gräf, Bettina. Medien-Fatwas@Yusuf Al-Qaradawi: Die Popularisierung des Islamischen Rechts. Berlin: Klaus Schwarz Verlag, 2010.

Gulf Times. Islamweb Enters 20th Year with New Look. Gulf Times. (2018-05-20). Consulted online on 02 May 2019. https://www.gulf-times.com/story/593452/Islamweb-enters-20th-year-with-new-look.

Hānī, Imān. داعية: ممارسة الجنس مع الروبوت صوفيا ليس زنا. Almesryoon. (2018-04-29). Consulted online on 30 September 2019. https://almesryoon.com/story/1167249/داعية-ممارسة-الجنس-مع-الروبوت-صوفيا-ليس-زنا.

Hawting, Gerald R. The Idea of Idolatry and the Emergence of Islam. Cambridge: Cambridge University Press, 1999.

Klein, Mechthild. "E-Priester im Einsatz." Deutschlandfunk. (2019-09-25). Consulted online on 26 September 2019. https://www.deutschlandfunk.de/religion-in-japan-e-priester-im-einsatz.886.de.html?dram:article_id=459483.

Kutscher, Jens. The Politics of Virtual Fatwa Counselling in the 21st Century. Masaryk University Journal of Law and Technology 3, no. 1 (2009): 33–49.

Noyes, James. The Politics of Iconoclasm. London: Tauris, 2013.

Pfotenhauer, Sebastian and Sheila Jasanoff. Traveling Imaginaries: The "Practice Turn" in Innovation Policy and the Global Circulation of Innovation Models. In The Routledge Handbook of the Political Economy of Science, edited by David Tyfield, 416–428. Routledge: London & New York, 2017.

Promta, Sompar and Kenneth Einar Himma. Artificial Intelligence in Buddhist Perspective. Journal of Information, Communication and Ethics in Society 6, no. 2 (2008): 172–187.

Tamatea, Laurence. Artificial Intelligence, Networks, and Spirituality: Online Buddhist and Christian Responses to Artificial Intelligence. Zygon 45, no. 4 (2010): 979–1002.

Tegmark, Max. Life 3.0: Being Human in the Age of Artificial Intelligence. New York: Alfred A. Knopf, 2017.

Tijani, Ismaila. Ethical and Legal Implications of Sex Robot - An Islamic Perspective. International Journal of Sustainable Development 6, no. 3. (2012): 19–27.

Tyan, Émile and James Richard Walsh. Fatwā. In Encyclopaedia of Islam, Second Edition, edited by Peri J. Bearman, Thierry Bianquis, Clifford Edmund Bosworth, Emeri van Donzel and Wolfhart P. Heinrichs. Consulted online on 02 May 2019. http://dx-1doi-1org-1ffotf59104ff.emedia1.bsb-muenchen.de/10.1163/1573-3912_islam_COM_0219.

Varvaloucas, Emma. Ai, Karma & Our Robot Future. Tricycle. (2018). Consulted online on 02 May 2019. https://tricycle.org/magazine/artificial-intelligence-karma-robot-future/.

Wensinck, A. J., and T. Fahd. Ṣūra. In Encyclopaedia of Islam, Second Edition, edited by Peri J. Bearman, Thierry Bianquis, Clifford Edmund Bosworth, Emeri van Donzel and Wolfhart P. Heinrichs. Consulted online on 02 May 2019. http://dx-1doi-1org-1ffotf5190719.emedia1.bsb-muenchen.de/10.1163/1573-3912_islam_COM_1125.

PART V

Conclusion

CHAPTER 13

Outlook for the Future of AI in the GCC

Elie Azar and Anthony N. Haddad

1 Contribution to the Literature

Artificial intelligence (AI) is arguably the most disruptive force technology for the coming decade (Bloomberg Intelligence 2017). In the wake of the fourth industrial revolution, businesses and governments are at an important crossroads: embrace the technological disruption or be left behind (PwC 2018). Acknowledging this new paradigm, GCC nations have made AI an integral element of their national strategies with concrete actions towards developing its potential and applications. Notable examples include the Saudi Vision 2030 national transformation strategy with digital transformation as a key enabler, a 2019 Saudi decree to establish a National Authority for Data and AI, a national UAE AI strategy and appointed Minister of AI, as well as other initiatives across key sectors, such as transportation and education. The above strategies and action plans are driven by projections of AI contributions to the economies of

E. Azar (✉)
Khalifa University of Science and Technology, Abu Dhabi, UAE
e-mail: elie.azar@ku.ac.ae

A. N. Haddad
Amazon, Dubai, UAE

E. Azar and A. N. Haddad (eds.), *Artificial Intelligence in the Gulf*,
https://doi.org/10.1007/978-981-16-0771-4_13

the Middle East estimated at US$320 Billion by 2030, mainly due to expected increases in productivity and benefits to consumers (PwC 2018).

Similar to other major technological innovations, the power and potential of AI come with great responsibility and risks on individuals, organizations, and society as whole (McKinsey & Company 2019), with the speed of innovation often outpacing the identification and treatment of these risk areas. Therefore, mapping AI's multifaceted challenges (and opportunities) for the region is of outmost importance to successfully harness the power of the technology while mitigating its potential risks and unintended consequences on society. Currently, the vast majority of books on AI in the literature detail its concepts and techniques (e.g. Russell and Norvig [2016]), applications to sectors such as business (e.g. Yao et al. [2018]) and education (e.g. Aoun [2018]), or the evolution of the sector and the resulting societal impacts (e.g. Tegmark [2017]). In parallel, few books have investigated the Digital North-South divide, with a focus on AI innovations from the Digital North and rivalry between the United States and China (e.g. Lee [2018] and Xiang [2019]), and without regard for the prospect of the Digital South's ability to contribute towards (as opposed to just adopt) AI technology and solutions. The Gulf region is especially understudied in the literature on the Digital South, despite the outsized interest from Gulf rulers and significant economic potential and growth projections for the region.

This edited volume directly filled the mentioned gap with unique contributions and case studies under three main themes: (1) Data, Governance, and Regulations; (2) Existing Opportunities and Sectoral Applications; and (3) Society, Utopia, and Dystopia. The chapters were adapted from papers presented in a workshop on AI in the GCC held at the University of Cambridge, UK, as part of the 2019 annual Gulf Research Meeting. The chapters benefited from extensive discussions with the participants of the workshop, including academics, practitioners, and policymakers of AI-related disciplines from the six GCC countries. The key takeaways from the different chapters are presented next, followed by a summary of the recurring themes and final thoughts.

2 Ten Takeaways

This section summarizes key findings from the ten core chapters of this book (Chapters 3-12), which exclude the introductory and concluding chapters.

Takeaway #1: Access to reliable data sources is one of the most important prerequisites for quality research and innovation in AI. The current processes to access public sector information remain unclear and unstructured, with concerns pertaining to data quality, security, and its potential misuse. Open data best practices can help mitigate the stated risks but require careful adaptation to the local context of GCC countries.

Takeaway #2: Stakeholder engagement is another condition for an effective and inclusive integration of AI in the different sectors of the economy. While top-down governmental efforts are needed and currently visible in most GCC countries, they should be met with bottom-up initiatives from the public sector, private sector, and civil society to create synergies that make the current interest and investment in AI sustainable.

Takeaway #3: The recent and fast-paced developments in AI systems call for regulatory interventions by policymakers that strike a balance between the expected benefits to society and potential threats and risks. While regulatory interventions often entail the implementation of "hard" and legally binding laws, incentive-based approaches, such as ones based on certification mechanisms, can offer fast and flexible risk-management solutions.

Takeaway #4: AI is a disruptive technology, especially for labour markets. While AI creates jobs, it also eliminates traditional jobs, creating an imbalance in workforce requirements. In general, there is a lack of research about the perceptions, attitudes, and knowledge of professionals towards AI in GCC institutions. Initial evidence points to an increasing number of people becoming "technophobic" with fears of their jobs being taken over by robots and other AI software. More research is needed on the topic, especially in the light of diverse nationalization programs of the GCC workforces.

Takeaway #5: AI investments in critical sectors, such as healthcare, should not be approached in isolation of regional and global efforts in the field. Global partnership and collaboration are needed to truly harness the power of the technology, particularly when tackling cross-boundary challenges that require collective actions, such as pandemics, economic crises, and climate change.

Takeaway #6: In parallel to global partnerships, local ecosystems should be adapted to provide enough flexibility and incentivize businesses to expand AI applications in their sectors. A possible venue to create such ecosystems is the concept of "free zones", which are geographic areas with the rules of business different from those of the national territory they are in. Free zones have played a vital role in cities like Dubai in attracting leading tech firms to the region. Consequently, free zones can play a similar role to create an AI-friendly ecosystem where AI technologies are developed and advanced locally, and to help in the region's transition from merely adopting to also developing AI solutions.

Takeaway #7: In parallel to the AI premise of increasing efficiency of systems and sectors, it is equally important to question how and whether AI is truly contributing to an improved well-being of the citizens. Similar questions are often asked when studying smart city technologies (e.g. IoT), reemphasizing the need for "socially smart cities". An important step in this direction is the "Happiness Agenda" of Dubai, developed under the "Smart Dubai" initiative.

Takeaway #8: Inclusiveness, particularly women's inclusion in AI, is a significant factor in ensuring the successful development, deployment, and governance of AI. As governments jockey for power in today's fourth industrial revolution, they need a workforce with the skills and abilities to match labour market demands and compete at a global stage. Therefore, national AI strategies should adopt "gender mainstreaming" that considers gender-equality at all levels of decision-making.

Takeaway #9: Tech companies are gathering and analysing a significant amount of data and information on their users, which, with the support of AI technology, make their personality traits, expense patterns, or even political beliefs become fairly predictable. The consequences of such knowledge result in significant ethical, health, and political implications that need to be studied and mitigated. Limited research on the topic is originating from the GCC, highlighting an important gap that needs to be filled to mitigate the stated negative effects of AI on Gulf citizens.

Takeaway #10: Given the social implications of AI, it is important to observe and understand its perception from the religion's perspective, particularly Islam. The study of Arabic and English Islamic legal opinions indicates that Islamic scholars have a fairly clear stance on the treatment of robotics, but not on AI. The scholars do not show clear concerns that AI technologies could harm humans or that their creators usurp God's power to create. Finally, they rarely cover difficult issues in their texts, such as the impacts of developing strong AI on society.

3 Recurring Themes and Final Thoughts

To conclude and summarize the recurrent themes in this book, a "Word-cloud" is shown in Fig. 1, which is a graphical representation of the most used words across the chapters. The size of each word in the figure is proportional to its frequency of occurrence.

As shown in Fig. 1, a common theme in the book is governance, represented by words such as "government", regulatory", and "organization", which is not surprising given the government-led AI initiatives that are currently being driven by the different GCC nations. On the other hand, terms pertaining to the private or educational sectors are less present in the figure, reconfirming the current dominance of top-down approaches to promoting AI in the region.

In terms of geographical representation, it is worth noting in the figure the size (i.e. occurrence) of the words "UAE" and "Dubai" in the manuscript, while other cities and countries are less represented. While the current chapters are not necessarily statistically representative of the larger literature on the topic, they still highlight what is a disproportionate

Fig. 1 Book "Wordcloud"

focus on activity in the UAE, suggesting a modest level of activity originating from some GCC states that may learn from the UAE's experience and leapfrog into the AI domain.

Finally, in parallel to "hard sciences" terms related to AI (e.g. "technology", "algorithm", and "robot"), "soft social" sciences terms are also present. Examples include "psychology", "humans", "employee", "inclusion", "cognitive", and "ethics". Such occurrences are very promising as they highlight a growing trend to approach AI with a human-in-the-loop perspective and to account for the social ramifications when adopting or implementing AI solutions. Ultimately, the role of technology is to serve humans and to improve their lives, contributing to their happiness and well-being.

References

Aoun, J., E. (2018). Robot-Proof: Higher Education in the Age of Artificial Intelligence. The MIT Press, Cambridge, MA, USA.

Bloomberg Intelligence (2017). A new era: Artificial intelligence is now the biggest tech disrupter. https://www.bloomberg.com/professional/blog/new-era-artificial-intelligence-now-biggest-tech-disrupter/.

Lee, K. F. (2018). AI superpowers: China, Silicon Valley, and the new world order. *Houghton Mifflin Harcourt*, Boston, MA, USA.

McKinsey & Company (2019). Confronting the risks of artificial intelligence. https://www.mckinsey.com/business-functions/mckinsey-analytics/our-insights/confronting-the-risks-of-artificial-intelligence#.

Russell, S. J., & Norvig, P. (2016). Artificial intelligence: A modern approach. *Pearson Education Limited*, England, UK.

PricewaterhouseCoopers (PwC) (2018). The potential impact of AI in the Middle East, *PwC*, London, UK.

Tegmark, M. (2017). Life 3.0: Being human in the age of artificial intelligence. Knopf.Life 3.0: Being Human in the Age of Artificial Intelligence. *Vintage*, New York, NY, USA.

Xiang, N. (2019). Red AI: Victories and Warnings From China's Rise In Artificial Intelligence. *Amazon Digital Services*, Seattle, WA, USA.

Yao, M., Zhou, A., & Jia, M. (2018). Applied artificial intelligence: A handbook for business leaders. Topbots Inc., New York, NY, USA.

Index

E. Azar and A. N. Haddad (eds.), *Artificial Intelligence in the Gulf*,
https://doi.org/10.1007/978-981-16-0771-4